THE AMERICAN COMBAT
AIRCRAFT
AND HELICOPTERS
OF THE
VIETNAM WAR

ENZO ANGELUCCI

THE AMERICAN COMBAT AIRCRAFT AND HELICOPTERS OF THE VIETNAM WAR

Illustrations by Pierluigi Pinto

PORTLAND HOUSE
New York

In this poster book the tables
are designed so that if you want
you can frame them.
For the single ones cut along the dashed line,
whereas the twin tables
can be pulled out of the book
by cutting the binding thread.

Copyright © 1986 ERVIN srl., Rome
English translation copyright © 1986 ERVIN srl., Rome.
All rights reserved.

This 1986 edition
published by Portland House,
distributed by Crown Publishers, Inc.,
225 Park Avenue South,
New York, New York 10003

Created by Adriano Zannino
Editorial assistant Serenella Genoese Zerbi
Editor: Maria Luisa Ficarra, Victoria Lee
Final texts and technical data by Paolo Matricardi
Translated from the Italian by John Gilbert

ISBN 0-517-61775-7

Colour separation SEBI srl., Milan
Typesetting Tipocrom srl., Rome

Printed in Italy by SAGDOS S.p.A., Milan

The Publisher would like to thank the official authorities
of US Air Force, US Navy and US Marine Corps for their cooperation

CONTENTS

THE REASONS OF A WAR

It is still too soon to put what is commonly known as the Vietnam War in its proper historical context; today, so short time after it ended, one can do little more than trace the outline of events.

This undeclared war, described by many as the 'dirty war,' was the biggest occurrence of its kind since the end of World War II, in terms of duration, of the number of men and materials used, of lives lost, both military and civilian, and of its consequences on the political face of a large part of the world.

It was a war in which the United States found itself involved because it remained loyal to its principle of acting as a defender of freedom and democracy, which has long been its cherished belief, and in this it was joined by nations such as Australia and New Zealand, whom nobody could possibly accuse of being imperialist or colonialist powers.

It was a war that cost the United States more than 50,000 dead, 300,000 wounded, thousands of millions of dollars, and unprecedented internal laceration at the hands of public opinion. It was a war that America fought, as it were, with one hand behind its back, in an attempt to minimize civilian casualties in the enemy camp, in the hope that a mere show of strength would be enough to persuade the enemy to back down, and in the determination to avoid at all costs the intervention of two other superpowers, the Soviet Union and China, which gave ideological and material support to the enemy. The active involvement of the United States and its Southeast Asia Treaty Organization (SEATO) allies was progressively stepped up only in response to an increasing Communist threat to South Vietnam. The United States and its allies were not prepared for the type of war imposed by geographical and local climatic conditions and by enemy tactics. They were ill equipped to fight a war against an invisible enemy which would pop up now here, now there, as if by magic. Although during the early 1960s, for example, the USAF considered itself reasonably well prepared to fight a nuclear war, with its tactical supersonic fighter-bombers and its gigantic aircraft specifically designed for carrying atomic bombs, it was completely unequipped for a war such as that in Vietnam, where the most valuable weapon was soon shown to be an anti-guerilla airplane, the COIN (Counter Insurgency), capable of landing and taking off from short, makeshift runways, and carrying a wide range of armaments. Strange as it may seem, this type of aircraft was not part of the American arsenal, and at the beginning it proved necessary to adapt old World War II planes such as the Skyraider or the T-28. It was indeed some time before the principle of «airmobility» was adopted; and when it was, in the course of the war, it gave the helicopter a role that during the 1950s would have been unthinkable.

From August 1964 to April 1973 the United States not only demonstrated its immense power on the battlefields of Vietnam but also gave clear proof of its superior technological capacity. As far as power goes, it is sufficient to point out that the American air forces were capable of unleashing on the enemy some 2.5 million tons of bombs (half a million tons more than in the entire course of World War II) and 19 million gallons of defoliants, yet doing negligible damage to an elusive, dispersed enemy and in no way eroding the latter's powers of attack and resistance. It is also evidence of its potential might to point out that the United States was able to lose about 10,000 planes and helicopters in the war without experiencing any difficulty in keeping its flight paths open. And it demonstrated the same capabilities when, quite soon after hostilities began, it launched new types of aircraft and helicopters, such as the excellent US Navy Douglas Invaders and Grumman Prowlers, the Rockwell Broncos, and the Bell HueyCobras, all better adapted for this specialized kind of war, not to mention the McDonnell Phantom II which was the most effective answer to the MiG-17s, 19s and 21s which were thrown into the fray by the North Vietnam air force. During hostilities, of the 184 MiGs shot down by the Americans, some 137 were hit by air-to-air Sidewinder and Sparrow missiles or by gunfire from USAF or US Navy Phantoms. The veteran Republic F-105s were credited with 23 successes and the US Navy Vought Crusaders with 15. The United States lost only 74 aircraft, as against 181 MiGs, in aerial duels, with a proportion of 3.86 in favor of the US Navy and 2.15 for the USAF. It was in Vietnam that the Phantom showed itself to be one of the best fighters and fighter-bombers in the world. The Phantom may also be regarded as the best example of the superior technology tested by the Americans in Vietnam, along with the Sidewinder and Sparrow air-to-air missiles, the Bullpup air-to-ground missiles, and the 'smart' bombs. In the course of the war, technology in this field made giant strides: after 1967 the US Navy, in addition to freefall bombs, made use of the «Walleye,» guided by videocamera in which the target had previously been framed; and from 1970 the USAF added the LGB (Laser-Guided Bomb). Thanks to the precision of these new devices, targets which had seemed untouchable, especially bridges, could now be hit with accuracy. The Hughes TOW, an anti-tank guided missile, which could also be launched from a helicopter and which is nowadays part of western arsenals, got its baptism of fire in Vietnam. Equally remarkable was the development of electronic warfare. To the initial launching by North Vietnam of SA-2 ground-to-air anti-aircraft missiles, of Russian construction, American technology responded immediately in 1965 both by introducing Wild Weasel aircraft, especially equipped for pinpointing and destroying the missile launching bases, and by allocating to other aircraft a special container with anti-SAM devices. In this respect the gap between Russian anti-aircraft missiles and American countermeasures narrowed rapidly, to the latter's advantage. At the end of the day, there were more victims of traditional anti-aircraft weapons than of the newfangled kind.

The Vietnam War, by reason of its characteristics and needs, also promoted a new form of mobility on the battlefield. World War II had witnessed astonishing and sudden frontal movements, thanks to the mobility afforded to the armies by tanks and tracked vehicles which made it possible to shift entire divisions rapidly; but in territory that consisted mainly of jungle, and against an enemy which did not deploy its forces along a front line in the proper sense, the only means of mobility, whether offensive or defensive, was afforded by the helicopter. This already became clear early in 1962 and resulted in the adoption of a new tactical pattern based on the use of transport helicopters. These were escorted by armed helicopters, protecting the vital moments when troops disembarked, often very near (and

sometimes in the midst of) enemy concentrations capable of delivering heavy fire power with rifles and automatic weapons. At first the idea of airmobility was considered applicable only to groups detailed for prompt intervention, known as «Eagle Flight,» composed of one Huey helicopter for the officer in command, seven helicopters carrying troops, five helicopters armed for close-range defense, and one ambulance-helicopter, escorted, when possible, by a pair of Skyraiders. It was only when, in October 1965, the 1st Cavalry Division, with its 400 helicopters and 1,600 vehicles arrived on the scene, that the concept of air mobility began to be applied on a wide scale, with immediate advantages on the battlefield. Such advantages would in due course be attenuated and nullified in face of simultaneous attacks by a number of Viet Cong formations. While the troops were engaged in one particular action, their presence would suddenly be required in at least five other sectors. The apparent lack of success of the US Army and Marines in Vietnam in no way detracts from their courage and their spirit of sacrifice; and a special tribute should be paid to those very youthful helicopter pilots, mostly drafted men, who, after sixteen weeks at Fort Wolters, Texas, where they learned to fly, and a short combat training period at Fort Rucker, Alabama, frequently found themselves caught up in an inferno of firepower during their very first missions.

Although the US Army paid the heaviest toll in terms of dead and wounded, the other armed forces also suffered cruel losses. The Marines, always sent in for the most perilous missions, lost 12,936 on the battlefield and had 88,594 men wounded; the figures for the USAF were 2,118 dead and 3,460 wounded, with 586 men captured or missing; and the US Navy lost more than 1,000 pilots, radar operators and helicopter pilots.

From October 31, 1968 to 1972, the USAF carried out over 233,000 war missions. Although most of the actions involved the men of the Tactical Air Force of the Pacific Air Force — used for Forward Air Controller (FAC), in support actions of ground forces, in attacks on objectives both in South and North Vietnam, and in Combat Air Patrol (CAP) — the Strategic Air Command (SAC) also played its part, with its B-52s and, above all, its KC-135 tankers. Mention must also be made of the support role of the Military Airlift Command (MAC), which probably had the most difficult and responsible task in its entire history, transporting hundreds of thousands of men and millions of tons of military material from the United States to Vietnam, and ferrying tens of thousands of wounded from Southeast Asia back home.

The US Navy likewise made an enormous contribution to the war, using all its aircraft carriers continuously in rotation.

It is clear, therefore, that all branches of the American armed forces were used to the hilt in Vietnam, and that all of them fought unsparingly in an undeclared war that the United States, right from the start, was destined, if not to lose, certainly not to win. I believe that today everyone, American or not, must agree with Vice Admiral Ralph W. Cousins, Commander of Task Force 77 in 1977, who declared: «I am certain that the United States has never fought a war in which our young men have been as courageous, as competent, as they have in this one.»

VIETNAM 1945-1975

At the end of World War II, the former thriving French colony in Indochina was abandoned by the defeated Japanese occupying troops and taken over in the south by British forces and in the north by those of Nationalist China. But on September 2, 1945, when the Chinese reached Hanoi, they found the territory north of the 16th parallel virtually under the control of Ho Chi Minh, the man who had been head of the resistance against Japan and the founder of the Vietnamese Communist movement known as the Viet-Minh.

Having no intention of accepting a return of French colonialism, Ho Chi Minh, on the very day of liberation, proclaimed his Declaration of Independence and the foundation of the Democratic Republic of Vietnam.

France, however, whose former colony was now officially returned to her, was in no way prepared to accept the fait accompli; from the final months of 1945, therefore, a full-scale encounter erupted between the Viet-Minh and French troops, who had immediately re-entered the colony. This conflict lasted until 1954, when the French were finally defeated.

From 1945 to 1954, thanks to French initiative, the whole of Indochina was divided into three autonomous states, all under French control, namely Vietnam, Laos and Cambodia, in an attempt to counter local independence and autonomous movements. In Vietnam Emperor Bao Dai assumed power, and he was recognized by the United States government on February 3, 1950.

Shortly before this, in December 1949, the Communist army of Mao Tse-tung had finally seized power in China, thus creating a situation on the northern border with Vietnam which was far more to the liking of Ho Chi Minh. Although the new Chinese Communist regime was not immediately in a position to help its ideological neighbour, its very existence acted as a spur and incentive to step up the struggle against the French and against colonialism, with the ultimate goal of independence. The United States was implicated for the first time in the future of Southeast Asia on February 16, 1950, when France was compelled to request help. The American colossus was thus forced to think seriously, as never before, about a distant corner of the world which only a very few of its citizens were even aware of, notably a handful of veterans of World War II who, as USAAF pilots from 1942 to 1944, had seen action against the Japanese at Hanoi, in the port of Haiphong, at Saigon and in the Tonkin Gulf. To lend the French massive support would have been simple and comparatively inexpensive, but it would have meant backing a colonialist nation, which it had itself condemned as unjust. There were those who began to advance the so-called «domino theory,» whereby if one piece were to fall all the others might follow suit, implying that were Indochina to be lost to the Communists, the same would happen to Thailand and Burma; but few were inclined to listen. The consequence was that on May 1, 1950 President Truman decided to tender France, for the defense of her colony against Communism, aid only to the tune of ten million dollars, and to send a very small number of American officials to Vietnam to set up the Military Assistance Advisory Group (MAAG) with the job of monitoring developments first-hand and giving more assistance to the puppet government of Bao Dai than to the French anti-guerilla activities. So it was President Truman, the man who had brought to an end World War II by ordering the atomic bomb to be dropped on Japan, who first involved the United States in the affairs of Vietnam.

From 1950 to 1954 there was fierce fighting between the forces of the Viet-Minh independence movement and those of the puppet governments backed by the French since 1949. It was a conflict with no quarter given, and it should have taught something to the American observers, making them understand that this was no ordinary war but a guerilla conflict in which the enemy was for most of the time invisible, except when actually on the attack. In the Franco-Vietnamese war the air forces played a negligible role, probably because the Viet-Minh lacked air support and the aircraft available to the French were nothing to boast about. L'Armée de l'Air had at its disposal a miscellany of airplanes which included captured Japanese Ki-43 fighters, Ki-55 attack aircraft, a few old Spitfires, Hurricanes, Hellcats, Bearcats, Mustangs, Mosquitos and B-24s and, for transport, the veteran German Ju-52/3ms. These were enough to give the French air supremacy of a sort but they did little to blunt the guerilla offensive. The unsuitable planes carried out attacks on scattered groups of guerillas, while larger enemy formations would appear from nowhere to ambush French convoys.

Left without any really significant aid from the British and Americans, the French, already short of manpower and material as a result of a tragic, exhausting world war, were finally defeated; on May 7, 1954, Dien Bien Phu, defended by 600 men of the legendary Foreign Legion who by now had neither munitions nor supplies, was overrun by Viet-Minh forces seventy times their number. This marked the end of French colonialism in Southeast Asia.

While Ngo Dinh Diem took over the post of Prime Minister in South Vietnam, all the major powers hoped that some form of balance would be achieved between the Communist forces of North Vietnam and those who were staunchly non-Communist in the prevalently Catholic South. In July 1954, at an International Conference held in Geneva, representatives of the great powers recognized the independence of Cambodia, Laos and Vietnam, the three nations to emerge, for ethnic reasons, from the division of Indochina.

Because it was split by factions, Vietnam was to hold nationwide free elections in July 1956 to decide on the type of government that most of the people wanted. Meanwhile the country would be divided at the 17th parallel by a Demilitarized Zone (DMZ), under the control of an International Commission. When these decisions were made public, almost a million non-Communist North Vietnamese resolved to emigrate south and 300,000 of these were evacuated by ships lent by the US Navy, responding to the appeal on humanitarian grounds. On the other hand, 100,000 residents of South Vietnam, principally Communists, left for the north where they subsequently formed the nucleus of the movement for liberating South Vietnam.

The United States hoped that the decision made under the Geneva Protocols would prove lasting, encouraged by the fact that the previous attempt at compromise in Korea had succeeded in stemming Communist infiltration into the south. On September 8, 1954, the US convened a conference in Manila, Philippines, which set up the Southeast Asia Treaty Organization (SEATO), consisting of the United States, the United Kingdom, France, Australia, New Zealand, Pakistan, the Philippines and Thailand. The treaty called upon each of the member states to defend any of the countries who might be subjected to serious Communist threat. This was to set the scene for subsequent events.

On October 5, 1954 the last French troops left Hanoi, and six days later, on October 11, Ho Chi Minh proclaimed for the second time the foundation of the Democratic Republic of Vietnam, while his supporters took firm control of all territory north of the 17th parallel. To pacify Diem, US President Dwight D. Eisenhower, on October 24, 1954, formally guaranteed that the Americans would not abandon their allies in South Vietnam. Truman's successor then, was to take a further step, albeit determined by circumstances, toward a more direct involvement by the United States in Vietnam.

The situation in either part of the country was very different. In the north tight discipline, political fanaticism and anti-Japanese and anti-French guerilla experience contributed very rapidly both to the creation of solid administrative machinery and an efficient military structure, even though poorly armed. In the south, government corruption, lack of discipline and a kind of native indolence made things difficult and the future even more uncertain. Despite this, on October 23, 1955, having declared his unavailability in national elections for fear of North Vietnamese intrigues, Ngo Dinh Diem deposed Emperor Bao Dai, proclaimed the Republic of Vietnam, and proceeded to establish an army and an air force. The first Vietnamese Air Force (VNAF) had been founded in 1951 under French sponsorship, with the setting up of a school for pilots at Nha Trang and the formation of the 1st Air Observation Squadron (AOS), equipped with Morane-Saulnier M.S.500s, the French version of the American L-19. The new VNAF, officially formed on July 1, 1955, was under the control of advisory-instructors of the Armée de l'Air. At the time of its reconstruction, the VNAF numbered some 1,350 men and 58 machines, mostly M.S.500s, Cessna L-19s, Grumman F8Fs, a few Douglas C-47s and some Junkers Ju-52/3m transport planes, almost all of which were based at Nha Trang, later renamed Da Nang.

Meanwhile, on May 10, 1955, Ngo Dinh Diem had officially requested the United States to send instructors for his armed forces. From April 28, 1956, therefore, coinciding with the departure of the last French units from the south, a new MAAG was formed in Saigon, detailed to concern itself immediately with the army, while the training of VNAF pilots was taken over by the Americans from 1957. In the meantime things had not been quiet on the Communist front; in May 1957 the Communist Pathet Lao movement had tried to seize power in Laos, and in January 1958 Viet Cong guerillas had the audacity to launch an attack on Saigon itself.

In May 1959 North Vietnam officially showed its hand: the Central Committee of the Communist Lao Dong party announced its intention of unifying the country even if it meant war. And while, in Laos, the Pathet Lao fought to win control of the northern part of the country, with the assistance of North Vietnam, the first 4,500 men of the North Vietnam Army (NVA), mainly consisting of South Vietnamese, began infiltrating south by way of a route through Laos which became famous under the name of the Ho Chi Minh Trail. In the south terrorism and guerilla activities increased from the latter half of 1957; so it became more than ever urgent that the South Vietnam army and air force should reach a degree of efficiency not only to deal with attacks from within but also to face the growing threat from the north. As a result of this, by the end of 1959 the strength of the MAAG went up from 342 to 685 military advisers, most of whom were assigned to the South Vietnam air force, which had one operative fighter squadron, equipped with ancient Grumman F8Fs, two observation squadrons with L-19s, one training squadron of T-6s, two transport squadrons with C-47s, and one helicopter squadron of Sikorsky H-19s, principally used for rescue operations.

In 1960, while Laos gradually fell into Communist hands, the National Liberation Front (NLF) emerged in South Vietnam, at a time when popular discontent with Diem's government was at its peak and when there was an unsuccessful coup d'état against the dictator.

In September 1960, at the insistence of the American military advisers, the United States decided to reinforce the VNAF by sending the first ones of 25 A-1H, formerly AD-6, Skyraiders from the US Navy, an aircraft which could somehow be used for antiguerilla missions, in place of the F8Fs which by now were not airworthy, and eleven Sikorsky H-34 Choctaw helicopters. This gave new life to the VNAF, even though few of these machines could be used continuously, either being grounded because of faulty maintenance or lack of spare parts.

In 1961 the Soviet Union officially declared its views on the Vietnam situation in an address by Nikita Khruschchev which announced full political support of North Vietnam. In response to this, US Vice President Lyndon B. Johnson visited Saigon in mid-May, and in the course of his stay confirmed, in President Kennedy's name, American intention not to abandon the Republic of Vietnam. In fact, following the report of the Vice President and that of special envoy General Maxwell B. Taylor, Kennedy decided on November 13 to step up military aid to Diem, handing over 30 North American T-28Ds (training planes adapted to play a COIN role), and sending three US Army squadrons of Piasecki H-21 Shawnee helicopters and one squadron, also from the army, consisting of sixteen Fairchild C-123 Providers. Even though these were transport rather than combat units, it was President Kennedy who dispatched the first American troops to Vietnam.

Just over a month later, the H-21s initiated the first airmobile operation in history, transporting 360 soldiers into an area infested by Viet Cong not far from Saigon, with an aerial escort of an L-19 observation plane and two Skyraider attack aircraft to support the action. The C-123s, however, were soon being used to spray defoliants over suspected jungle zones, in an attempt to flush out enemy formations. The first American plane, with an American crew of three men, to be lost in Vietnam was a Provider, shot down by anti-aircraft batteries while engaged in this mission. The Viet Cong, strengthened by the solidarity and complicity of the civil population, seemed to be at once everywhere and nowhere, and during this phase of the conflict enormous forces were employed to obtain derisory results. Roy M. Braybrook, in an essay which appeared in «Military Airpower,» remarked that according to reliable estimates, the death of a single Viet Cong cost, on average, seven tons of fuel, seventeen tons of munitions and one ton of napalm. Whether or not this was quite accurate, it was surely not far removed from the actual state of affairs.

Meanwhile, in September, a detachment of the 507th Tactical Control Group had arrived at the Tan Son Nhut base not far from Saigon, reinforcing a Control and Reporting Post (CRP) with a powerful radar and four McDonnell RF-101 Voodoo reconnaissance planes. The latter would later be replaced by another four RF-101s from the 45th Tactical Reconnaissance Squadron based at Don Muang Airport in Thailand.

In October men and material for the «Farm Gate» operation reached the Bien Hoa base. Back in April 1961, the 4400th Combat Crew Training Squadron, known as «Jungle Jim,» had been set up in Florida, gathering together the personnel to be transferred to Vietnam to prepare the young South Vietnamese pilots for combat. The 2nd Detachment of the 4400th, with 151 men, eight T-28Ds, four RB-26s and four SC-47s, was based at Bien Hoa. The Americans were intended simply to act as instructors but were ordered to return enemy fire if attacked, provided they were absolutely certain not to harm civilians. The Viet Cong, however, did not wear uniforms and were dressed like ordinary peasants; so it was almost impossible to determine whether or not they were Viet Cong, risking them having the advantage of striking the first blow. During the war, the inability to distinguish soldiers from harmless civilians was to cause the death of countless innocent people.

On February 8, 1962 the MAAG, by now numbering about 4,000 men, was transformed into the MACV (US Military Assistance Command, Vietnam), while the Viet Cong intensified their attacks on trains, military convoys, government positions and villages. In the spring of 1962 the 2nd Fighter Squadron of the VNAF, with T-28Ds, became operational, but its frenzied activity was insufficient to worry the enemy appreciably. The T-28s and Skyraiders, furthermore, were not equipped with systems enabling them to identify night targets, so SC-47s began to be used which could light up the zone under attack with flares. During that time 520 radios were distributed to the existent military posts scattered throughout the country. When a Viet Cong attack was confirmed, radio messages were sent requesting the intervention first of the SC-47s and then of the Skyraiders. But all this was theoretical, for in practice the handful of planes belonging to the VNAF were too busy elsewhere to be capable of answering any such request for help.

One other episode certainly did not help to sustain the spirits of the VNAF; on February 27, 1962 several A-1Hs attacked Diem's palace residence in a coup d'état which failed. There followed a period of purges during which the entire air force was regarded by Diem and his colleagues with suspicion, in spite of prodigious services rendered.

The Communist threat was daily more evident, and the agreements made by SEATO now came into effect; in August 1962 the first Australian contingent of the Military Aid Force (MAF) arrived in Vietnam.

In March 1962, the Americans had rendered operative the TACS (Tactical Air Control System) with a complex radar network. The sighting of two unidentified aircraft flying low over the highlands of South Vietnam led to the sending of four Convair F-102s to the area, with the aim of establishing a defense force in the event of North Vietnamese air attacks. Operation «Farm Gate» was itself strengthened with five more T-28s, ten B-26s and two C-47s, but all this represented token rather than substantive support, given the gravity of the situation.

Throughout 1963 the Viet Cong forces chalked up one success after another, while discontent among the South Vietnamese population increased by leaps and bounds as a result of the government's overriding power and evident corruption. On June 11 a Buddhist monk set himself on fire in a Saigon street as a protest against the regime's oppression, and others followed his example. The United States decided that it could no longer continue supporting such a dictatorial and corrupt government and abandoned Diem to his fate. On November 2, 1963 he was assassinated, and five days later General Duong Van Minh assumed power. Back home, meanwhile, public opinion expressed its growing concern and dismay at what was happening in Vietnam, and on November 15 Secretary of Defence McNamara, although reiterating American support for Vietnam in terms of arms and munitions, announced the withdrawal of the first 1,000 American advisers, whose numbers had by now exceeded 15,000,

and warned that by the end of 1965 the South Vietnamese would have to fend for themselves.

The unexpected departure of President Kennedy, assassinated in Dallas on November 22, left his successor, Lyndon B. Johnson, with a situation that was daily becoming more difficult. At that point the USAF had in Vietnam 117 aircraft, and the US Army, in addition to 325 fixed-wing planes and helicopters, had about 16,000 men. While Johnson, pressed by hawks and doves, declared he would make his own decisions, a new state of affairs suddenly arose in Vietnam: on January 30, 1964 Duong Van Minh was deposed. Power was assumed by General Nguyen Khanh, regarded as a man capable of lifting the morale and fighting spirit of the army and air force units.

Viet Cong pressure, in the meantime, had not diminished, and in May 1964 McNamara, though still proclaiming the intention of America to withdraw its forces quite soon, could do no less, given the seriousness of the situation, than assign more two-seater Skyraiders (A-1E) to the American Air Command, and single-seater A-1Hs to the VNAF. A month later, President Johnson nominated General William C. Westmoreland to the command of the MACV, and in July General Maxwell D. Taylor was made ambassador to Saigon. In the United States there were conflicting views: the hawks in the presidential circle recommended punitive raids against North Vietnam, while the doves urged the total withdrawal of American advisers, even though this might mean letting the whole country fall into Communist hands. Johnson, who had personally visited Vietnam, was by and large inclined to maintain the level of previous American assistance, convinced of the validity of the domino theory. He was awaiting the first reports from Westmoreland and Taylor when the Tonkin Gulf incident occurred.

On August 2, 1964 several North Vietnamese torpedo boats attacked the American destroyer USS Maddox in international waters. On the night of the 3rd and 4th, the same ship and the American destroyer USS C. Turner Joy signaled a further attack by the small enemy fleet. American reaction was immediate: on the same day four Vought F-8E Crusaders of the VF-53 Squadron aboard the aircraft carrier USS Ticonderoga, stationed in the area, attacked and sank, with missiles and cannons, one torpedo boat. Another 64 planes from the USS Ticonderoga and the Constellation struck at the North Vietnamese torpedo boat bases at Hon Gai and Loc Chao, hitting 29 vessels, and destroying fuel depots at Vinh, with the loss of three machines to anti-aircraft fire. The undeclared war against the Democratic Republic of Vietnam had begun. Feelings ran high at home; if a prompt response to the incident was not forthcoming, it might prove the case of those in Asia who described the United States as a «paper tiger.» On August 7, 1964 Congress, with the Tonkin Gulf Resolution, authorized President Johnson to take all necessary measures to prevent further acts of aggression and to respond with all available means to further North Vietnamese attacks. The first immediate step taken by the President was to dispatch to Vietnam two squadrons of Martin B-57s, one squadron of F-100 Super Sabre tactical fighters, and one squadron of Convair F-102 Delta Dagger fighter-interceptors, based at Da Nang, and other units based elsewhere. Thus did he initiate the 'escalation' which was to lead to the full and direct involvement of the United States in the war, even though air operations against North Vietnam would always be conditioned by many doubts, restrictions and hesitations on the Pentagon's part. On November 1, 1964, the same day as Tran Van Huong became Prime Minister of South Vietnam in place of Nguyen Khanh, Viet Cong guerillas delivered a mortar attack on the Bien Hoa base, destroying five Martin B-57s and damaging fifteen, as well as four VNAF A-1Hs.

One of Johnson's first decisions was to approve a program of attacks on the communication lines in Laos, the Ho Chi Minh Trail, along which Viet Cong munitions, supplies and reinforcements were finding their way into South Vietnam. In the interests of greater efficiency, the function of FAC (Forward Air Controller) was entrusted to the USAF instead of the US Army. In 1965 all the Army's fixed-wing observer aircraft were handed over to the USAF, so that any request for the intervention of planes in support of ground troops was no longer handled by a linking officer some distance away with the troops but by aircraft which patrolled their assigned zones continuously, signaling any suspicious movement below. The US Army Cessna L-19s were given the new name of O-1 Bird Dog, and were flown by experienced officers who, having spotted the enemy and called up the attack force, would then point out the target with smoke flares and remain in the area to adjust fire and monitor damage until the action was over. Compelled to fly slowly at low level, the FAC aircraft were excellent targets not only for standard anti-aircraft batteries but also for riflemen; yet such missions, when successful, were invaluable in flushing out groups of guerillas concealed in the jungle and in pinpointing enemy convoys on the terminal stretches of the Ho Chi Minh Trail.

Undeterred by the increased American activity, the Viet Cong even summoned up the nerve to attack, on Christmas Eve, 1964, the Brink Hotel in Saigon, killing and wounding many American advisers, and, on December 27, launching their entire 9th Division in an assault on the village of Binh Gia, southeast of Saigon. After a six-day battle, the government forces were defeated. On February 7, 1965 the Viet Cong attacked the Pleiku base, and this proved the last straw.

President Johnson ordered the first aerial reprisal on North Vietnam with Operation «Flaming Dart I.» The monsoon prevented USAF planes from taking part in the action, but planes of the US Navy's Task Force 77, stationed out in the Tonkin Gulf, 75 miles (120 km) from the coast of North Vietnam, the so-called Yankee Station, were heavily involved. The Task Force consisted of 125 ships with 64,000 men on board, and it included the aircraft carriers USS Coral Sea, Hancock and Ranger from whose decks planes struck at the port and barracks of Dong Hoi, north of the Demilitarized Zone. The monsoon kept the USAF planes grounded for the second reprisal operation, «Flaming Dart II» after a Viet Cong attack on Qui Nhon, but 99 aircraft from Task Force 77 made a massive raid on the barracks at Chanh Hoa.

The continual attacks north of the Demilitarized Zone had begun on March 2, 1965 with Operation «Rolling Thunder», intended to strike at objectives ever farther north, carrying the threat to the environs of Hanoi itself. The Pentagon was convinced that such operations would force the North Vietnamese to slow down their pressure on the south. However, they were unable to hit built-up areas, whatever the size, for fear of causing civilian casualties; nor could they attack radar stations and airports in case they killed or injured Russian or Chinese military advisers.» Consequently the USAF and US Navy planes had to restrict themselves to barracks, military vehicles once clearly identified, and scattered groups of soldiers. It was like using a gun to swat a mosquito. Very soon, however, permitted targets came to include communication junctions and brid-

ges, in an attempt to slow down the flow of fresh supplies coming overland from China to the northern cities and of reinforcements to the Viet Cong in the south. The first action of the USAF in the context of «Rolling Thunder» took place on March 2 itself, when 44 F-105s, 40 F-100s, 7 RF-101s and 20 B-57s, refueled in flight by KC-135s of the SAC, attacked Xom Bong. Even so, it is doubtful whether the results obtained were proportionate to the gigantic forces employed.

The primary objective was still the jungle that concealed the paths of the Ho Chi Minh Trail, both in Vietnam and in Laos, over which came supplies for the south. In the month of April 1965 alone the USAF carried out more than 1,000 attack missions on the Trail, hitting some transport vehicles, but it was like trying to find a needle in a haystack. Up to 1967, until the entire zone was protected by a dense network of anti-aircraft emplacements, the Bird Dogs of the FAC had to find the enemy «on sight.» Later, when the use of observer planes proved too risky, experiments were carried out as well with a system of acoustic sensors known as Adsid, Acuobuoy and Spikebuoy, which, when activated by the noise of engines or the sound of moving troops, transmitted alarm signals to an aircraft stationed in the zone, generally a Lockheed EC-121R, which proceeded to retransmit them to the operational center so that attack aircraft could be sent to the scene.

While between March and May 1965 the 9th Marine Expeditionary Brigade of Task Force 76 and the 173rd Airborne Brigade both arrived in Vietnam, following the first South Korean units, 2,000 strong, which had landed at the beginning of the year, the air forces continued their attacks both in the north (mainly US Navy planes), along the Ho Chi Minh Trail and on Viet Cong units in the south (mainly USAF planes). The US Navy, too, was organized to furnish tactical support to the ground forces in South Vietnam, establishing another base, about 100 miles (160 km) southeast of Cam Ranh Bay, called Dixie Station, accommodating the aircraft carrier SS Oriskany, which carried the VMF-212, the only Crusader squadron of the Marines. During 1965, however, some ten aircraft carriers used Yankee Station in rotation; and in the course of that year the US Navy carried out 30,933 missions in the north and 25,895 missions in the south, losing 100 planes and 82 pilots, and destroying over 800 trucks and 650 railroad cars.

From June 1965 the Republic of Vietnam had a new Prime Minister in the person of Nguyen Cao Ky, and the United States began to use its gigantic Boeing B-52s, flying from bases on Guam, to bombard the Trail, but carrying out their first mission against the north only on April 12, 1966. Operation «Rolling Thunder» continued uninterruptedly; in September the USAF alone carried out 4,000 attack missions. By the end of 1965 the total number of American soldiers and fliers in Vietnam was 181,000.

In the hope that bombardment of the north might dissuade Ho Chi Minh from pressing his attacks on the south, the United States, as a gesture of goodwill, suspended bombing north of the 17th parallel on Christmas Eve, but already by the end of January the Pentagon realized that the North Vietnamese were more determined than ever and bombardments were resumed. In 1965, meanwhile, in addition to installing more anti-aircraft guns and of increased caliber, the North Vietnamese had set up their first SAM ground-to-air missiles, of the Russian SA-2 type. The first missile base was discovered on April 5, 1965 fifteen miles (24 km) southeast of Hanoi, and the first victim of a SAM was a US Navy F-4C. In 1965 North Vietnam launched

180 ground-to-air missiles from 56 identified bases, but only eleven hit American planes, shooting them down. Although at the start the missile bases were not included in the list of permitted targets, for fear of killing Russian military advisers, authorization was soon given. It was during 1965, too, that the first MiGs made their appearance in the skies over Vietnam. The air war proper began, in fact, in April when two F-105s, loaded with bombs, were shot down by a MiG during a mission against the north. Before the end of the year, however, over 60 MiG-15s and MiG-17s were shot down, though their airstrips were regarded as prohibited targets until 1967.

While the Phantoms of Combat Air Patrol (CAP) took on the MiGs, the Douglas EB-66Cs and Republic F-105F Wild Weasels of the USAF, and the Douglas EA-3 Skywarriors and Grumman A-6 Intruders of the US Navy concentrated on the SAM launch bases, finding them and destroying them. Almost all the American combat planes, therefore, were furnished with a special container for electronic countermeasures, so as to minimize the effects of this new weapon which received its baptism of fire in Vietnam.

The USAF, in 1965, introduced an excellent radar system, based on ground stations, ships and a Lockheed EC-121D which, in its turn, patrolled the skies over Saigon and was able to give warning of any suspicious movements to the Phantoms of the air defense CAPs. The first MiG-17 to be shot down was on June 17, 1965 when an F-4 of Squadron VF-21 from the aircraft carrier USS Midway, piloted by Commander Louis C. Page, encountered enemy fighters, hitting two of them with Sparrow missiles. The first victory for an F-4 of the USAF occurred on the following July 10, when two Phantoms from a patrol of four each shot down a MiG-17 with Sidewinder missiles. The F-4 pilots were Captains Kenneth Holcome and Thomas Roberts.

There was no letup in the fighting during 1966, with zones lost and regained, and fortunes changing. By the end of the year there were already more than 385,000 Americans in Vietnam. On the ground the 1st Cavalry Division (Airmobile), which had arrived in October 1965, together with the 101st Airborne Division and the 1st Infantry Division, deploying a considerable number of helicopters, were confronted with the grim reality of fighting an enemy which was virtually invisible, except when appearing from ambush. In the air the USAF continued to strike at enemy groupings north and south of the 17th parallel; and the US Navy concentrated its efforts mainly on power stations, bridges and lines of communication. Particular attention was given to the areas surrounding the port of Haiphong, into which flowed most of the supplies sent from the Soviet Union. However, the Navy had no authorization to attack the port itself, its unloaded vessels or the enormous quantity of cargo lined up on the wharfs, still as a result of Pentagon fears of Russian or Chinese involvement. So the Navy pilots had to go for such material only when loaded on trucks, which traveled at night or through the jungle, exposing themselves to SAM and anti-aircraft fire, when in theory they could have finished the matter in one fell swoop. This was really fighting a war with a hand behind one's back.

In 1966, however, the US Navy carried out some 30,000 missions in the north and some 20,000 in the south, losing 120 machines. At the beginning of 1967 the list of permitted objectives was expanded to include airports and even installations on the suburb of Hanoi which, in the meantime, had become the most strongly defended city in the world. The SAM emplacements had multiplied astonishingly and traditional anti-aircraft fire was proving ever more accurate and dangerous. Many American aircraft were hit and shot down in action over Hanoi and other parts of the north. It was only thanks to the bravery and dedication of the rescue and recovery crews that many pilots were picked up behind enemy lines by SH-3 Marine helicopters, escorted by Skyraiders which did their best to give covering fire in the teeth of fierce enemy attack.

According to a reliable source, during the first stage of the war, up to October 1966, of the 269 pilots and radar operators of the USAF and US Navy shot down over North Vietnam, 103 were rescued. Of those remaining, 75 lost their life crashing with their aircraft, 46 were taken prisoner, and 45 were missing without trace.

Again the principal American air targets were all the communication lines around the port of Haiphong, and especially the bridges, which were destroyed in hundreds by the US Navy pilots who, for the first time, were able to use the new Walleye bomb, guided to the target by a television system which guaranteed extreme accuracy of fire.

While the ground operations proceeded without any decisive victories, the political situation in Vietnam was more than ever confused; in September 1967 there was another power change at the top when General Nguyen Van Thieu was elected President of the Republic and General Nguyen Cao Ky Vice President. By this time there were 486,000 American servicemen in Vietnam, but this seemed to make very little impression on the North Vietnamese or the Viet Cong. In response to a further gesture of goodwill by President Johnson, who had decreed a halt to the bombing of North Vietnam over Christmas, the enemy began 1968 by launching what became known as the «Tet Offensive,» coinciding with the Buddhist new year. In addition to many strategic points in the highlands and farther south, one of the principal objectives was the American base of Khe Sanh, garrisoned by Marines, which was closely besieged by a number of North Vietnamese divisions.

The men under siege from January 22 to April 7 often found themselves in desperate straits, and only the strenuous efforts of the air forces stemmed the tide sufficiently to prolong resistance and avoid a defeat comparable to that of Diem Bien Fu. Helicopters of the Army and Marines, Provider and Hercules brought in supplies to the base at incredible risk to themselves; and planes of the USAF and US Navy, weather permitting, provided continual air cover of the perimeter, attacking the siege lines incessantly. Even the giant B-52s were called in to help break the ring, and sometimes they scored hits on the enemy only 328 ft (100 meters) or so from the Marine lines. Without any doubt, and in no way belittling the courage of those under siege, Khe Sanh held out only thanks to the miracles that came from the sky.

Apart from their presence at Khe Sanh, the air forces continued, in so far as the monsoon allowed, to strike with utmost vigor at objectives in the north. Only in January 1968 did the VA-153 and VA-155 Squadrons from the USS Coral Sea have strong reason to believe that they had finally put out of action, with Walleye bombs the Thanh Hoa bridge on the Ma river, 80 miles (129 km) from Hanoi. The bridge, a strategic objective of prime importance, had been attacked some 615 times by USAF and Navy aircraft since 1965; it had been bombarded with 1,250 tons of bombs, but had never suffered any damage that could not be repaired within a day or so. The Navy pilots were convinced, on January 28, that they had finished it off;

but on February 8 traffic began to move once more across the seemingly invulnerable bridge.

Meanwhile the situation on the ground was becoming increasingly critical: the Khe Sanh siege had been lifted, but how much longer could it have held out anyway, surrounded as it was by vast zones under complete enemy control? On May 31, 1968, President Johnson, having decided to resort to diplomatic channels to relieve the situation, ordered the interruption of bombardments north of the 20th parallel, and on October 31 the cessation of all offensive air operations north of the old Demilitarized Zone. So ended «Rolling Thunder,» after 304,000 tactical missions, 2,380 of which had been carried out by Boeing B-52s, in the course of which 643,000 tons of bombs were dropped on North Vietnam. Yet the operation had failed in its purpose of breaking the enemy's psychological resistance and exhausting his war potential. From the psychological viewpoint the North Vietnamese were firmly, perhaps blindly, determined to defend the cause of their country's unification and, dispensing forever with any type of colonialism, see their form of social justice prevail. From the military point of view, the absence of heavy industry, the security afforded to the supply access routes, the dispersed movement of men and material, the natural lie of the land, and the advent of the monsoon which gave days of breathing space for reorganization after a year of tension, had greatly diminished the effects of bombing. A single raid on the port of Haiphong or a naval blockade could have abruptly stemmed the flow of those supplies which were indispensable to Hanoi for the continuation of the war. But this was not on the cards, and it is impossible to say now, with hindsight, whether from a political viewpoint this was the only decision capable of preventing direct Chinese or Russian intervention. Militarily, it must be said that it is much more difficult to win a war that one is not genuinely determined to wage.

On May 3, 1968, President Johnson, under pressure from continuing antiwar demonstrations, accepted the North Vietnamese offer to open preliminary peace negotiations in Paris, and diplomatic meetings now began. In November Nixon was elected President of a nation wearied by a war that it could not win and which had by now drawn 536,100 young Americans into action. During 1969 peace negotiations got under way in Paris, even though Communist attacks on the south had not ceased. On June 5 American planes bombed the north again, as reprisal for the shooting down of a reconnaissance plane; but the withdrawal of troops of the US Army had commenced. The death of Ho Chi Minh, on September 4, appeared to have strengthened the North Vietnamese in their resolve to complete what they had started, especially at a time when final victory seemed within their grasp. By the end of the year the American presence in Vietnam was already down to 474,000 men.

In 1970 South Vietnamese and American troops attacked Communist bases on the border with Cambodia, but it was already obvious that the game was up. In December American numbers were reduced by a further 138,000 men, and by the middle of 1971 Australia, New Zealand and South Korea had also decided to withdraw their contingents.

At the end of 1971, from December 26 to 30, the United States resumed bombardments on the north yet once more as a means of putting pressure upon the North Vietnamese negotiators, and to remind Hanoi, still in aggressive mood, that American participation was not yet over. But this had no effect: on March 30, 1972 the regular army of the Democratic Republic of Vietnam officially invaded the territory of the Republic of Vietnam. On April 6, while the North Vietnamese swept on, meeting feeble resistance, Nixon ordered the resumption of bombardments of North Vietnam, and this time the USAF used laser-guided bombs. The famous Thanh Hoa bridge was destroyed on May 13 by 25 'smart' bombs of 2,000 and 3,000 pounds (907 and 1,365 kg) and 50 conventional bombs. It seemed to represent the crumbling of a symbol, but it was, in fact, American pride that lay in ruins.

On August 12 the last ground troops left Vietnam; only 43,500 air force personnel remained behind. In Operation «Linebacker I,» which lasted until October, American planes inflicted heavier damage on the enemy than in the three previous years of bombing. But diplomatic negotiations were going on and incursions were halted, to be resumed, however, on December 18 when the North Vietnamese delegation walked out of the Paris conference. Strikes against the enemy were now delivered with far less scruples than before, and were so massive that in early January the meetings were resumed, and on January 27 an agreement was reached which heralded the end of hostilities between the Democratic Republic of Vietnam and the United States of America. On March 29, 1973 the last American troops left Vietnam, setting the final seal on the only war that the United States had ever failed to win.

On April 30, 1975 Saigon fell, and by the end of the year Laos and Cambodia were also wholly in Communist hands. A united Vietnam finally reached, «liberty» came with purges, re-education camps and desperate flights toward death on board junks overloaded with people that nobody was prepared to accept.

For the United States Vietnam represented a sad page in history, but not one to be ashamed of, since it involved so much glory, bravery and sacrifice. This is even truer today when this great nation has shown itself ready at last to put everything into its right perspective, saluting in honor the men who, often at the cost of their life and of deep wounds both to body and mind, did their duty, and more, for their country.

Scale view of aircraft and helicopters

Cessna O-1F Bird Dog

Cessna A-37B Dragonfly

North American T-28D Trojan

Rockwell OV-10A Bronco

Grumman A-6A Intruder

Douglas A-1H Skyraider

North American F-100D Super Sabre

Douglas A-4F Skyhawk

Vought A-7D Corsair II

Northrop F-5A Freedom Fighter

Bell UH-1B Huey

Lockheed F-104C Starfighter

Bell AH-1G HueyCobra

Vought F-8J Crusader

McDonnell F-4B Phantom II

McDonnell F-4C Phantom II

14

Douglas AC-47 Gunship

Republic F-105D Thunderchief

Martin B-57B Canberra

Convair F-102A Delta Dagger

General Dynamics F-111A

Sikorsky HH-53C Super Jolly

Boeing Vertol ACH-47A Chinook

Boeing B-52 Stratofortress

NORTH AMERICAN   T-28D TROJAN

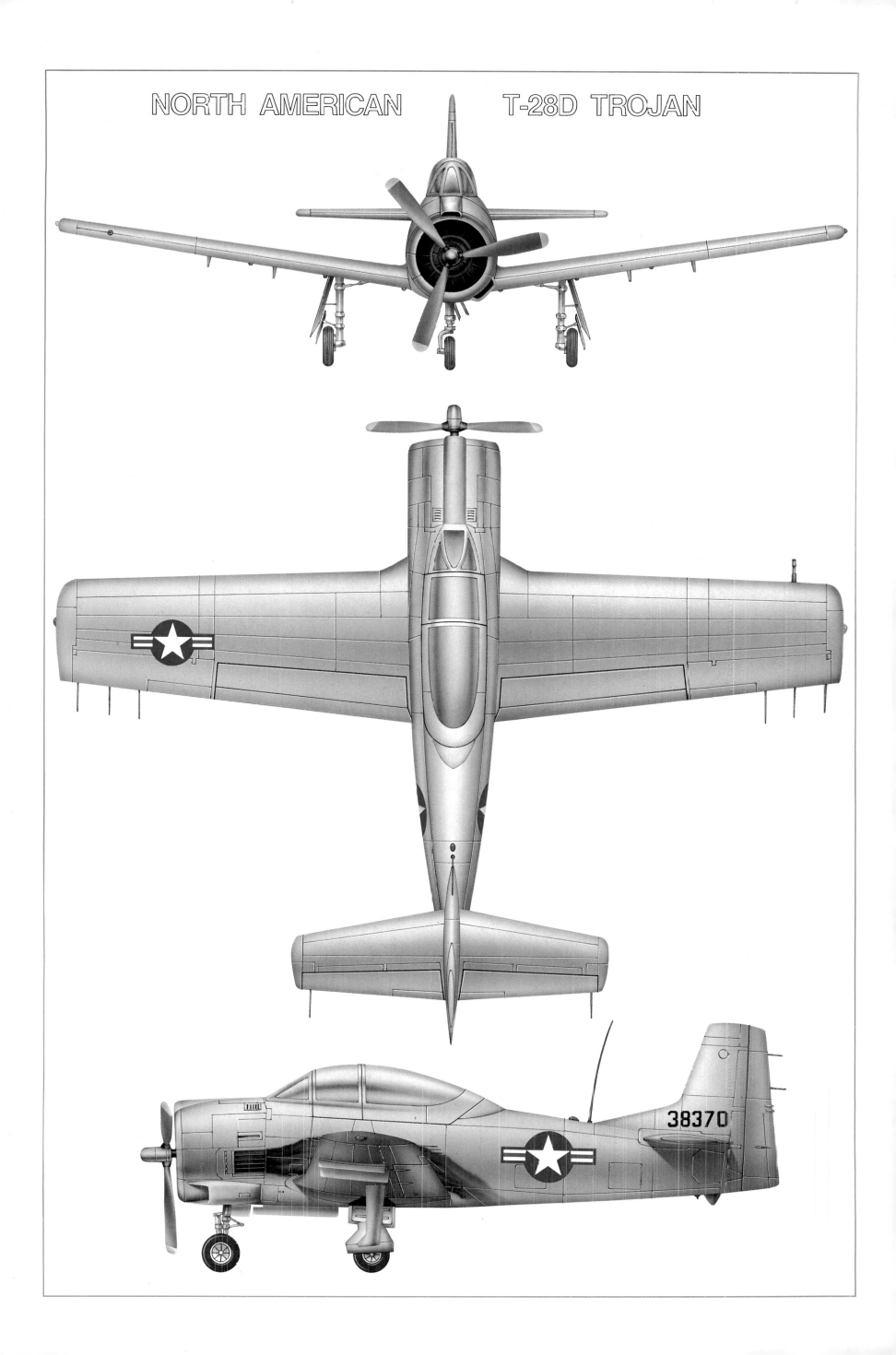

North American T-28D Trojan

The first eight T-28 Trojans reached Vietnam in March 1958, being handed over by the USA to the small South Vietnamese air force, which had been formed officially on July 1, 1955 as successor to the collaborationist air force set up by the French during the last phase of their stay in Southeast Asia. The planes were used for training pilots, as were the other 30 delivered to the VNAF at the same time as the first USAF unit to arrive in Vietnam, namely the 4400th CCTS (Combat Crew Training Squadron), which from October 20, 1961 was given the job of teaching the new South Vietnamese recruits. The 4400th Squadron was equipped with eight T-28s, four SC-47s and four RB-26s. But in 1962 Viet Cong pressure within the country began to make itself felt strongly, with ambushes on government columns and surprise attacks on villages and garrisons, so much so that it was considered indispensable to have aerial support ready to give assistance to the ground forces. At that point the T-28, because of the absence of true COIN (Counter Insurgency) aircraft, was transformed into the T-28D, with wing attachments for offensive purposes, in the form of bombs and rockets. In March 1962 the VNAF received another 30 of these; all were employed in action, sometimes even with a crew consisting of a South Vietnamese trainer pilot and an American instructor. Nevertheless the old and trusty Trojans proved well up to the new task and constituted the backbone of the strike forces of the VNAF up to 1964, when they were replaced by the Douglas A-1H Skyraiders.

Convair F-102A Delta Dagger

Given the possibility of the Chinese or Soviet air forces becoming involved in the confused situation that had come about in Southeast Asia, the first four Convair (then General Dynamics) F-102A Delta Daggers, at that time the best available American fighter-interceptor aircraft were sent to Vietnam, February 4, 1962. In August 1964, after the Gulf of Tonkin crisis, 12 further F-102As from the 405th Fighter Interceptor Squadron were dispatched to Da Nang and immediately used for CAP (Combat Air Patrol) cuties, designed to warn against and subsequently prevent North Vietnamese aerial incursions into South Vietnamese territory. However, they never got the chance to be engaged in combat, nor could they be used for other missions, being armed exclusively with air-to-air missiles. The same thing happened to the F-102s of the 509th FIS, sent to Vietnam in 1967. From 1968 the Delta Daggers were gradually called back to the US and replaced in their functions by the new Phantom F-4Es. Although stationed in the operations zone, the F-102s lost their unique opportunity to prove themselves in action because they encountered no enemy units whatsoever.

CONVAIR F-102A DELTA DAGGER

DOUGLAS   A-1H SKYRAIDER

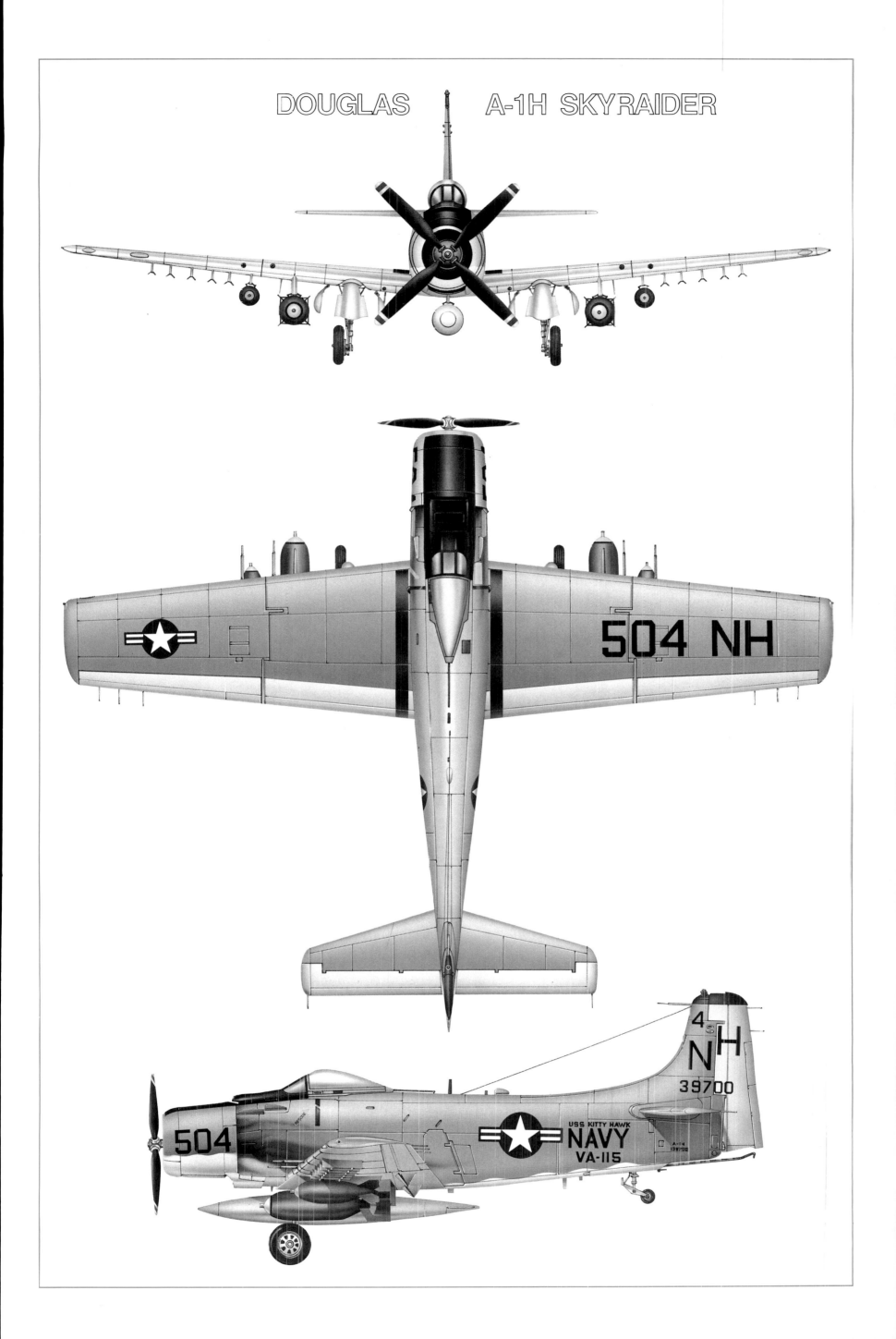

Douglas A-1E/H Skyraider

A piston-engined aircraft designed during the last years of World War Two, and a veteran of Korea, where it covered itself with glory, the Skyraider played a key role in the Vietnamese conflict, especially in the early years. Powerfully armed, slower than a jet aircraft, it came the closest to having the same characteristics as the anti-guerilla COIN, which at that time was desperately in demand. The A-1Hs were originally designated the single-seat AD-6s, and the A-1Es the two-seat AD-5s, last generation of the Skyraiders. At the time of the Gulf of Tonkin crisis, in August 1964, all aircraft carriers of the 77th Task Force had attack squadrons of Douglas A-1Hs, which were among the first sent into action against targets in North Vietnam. Two of them even achieved the incredible feat of shooting down two jet-engined MiG-17s. The US Navy Skyraiders remained at the front until April 1968, earning the nickname of «Workhorse of the Fleet,» and losing 48 machines, mainly from anti-aircraft fire. A number of aircraft of this type (25) surplus to Navy requirements, were also assigned to the VNAF, the first six in September 1960, the other ones in March 1961, as replacements for the now-obsolete T-28s. But even the USAF sent a few twin-seater A-1Es to the front in summer 1963, when the 1st Air Command Squadron was formed. Useful for low-level attacks, and ideal as escorts for rescue missions, during the early years of the war they did everything and flew everywhere, being given the familiar name of «Spad,» from the famous World War One fighter. Skyraider, the last piston-engined fighter aircraft of the US Navy, carried out prior to its final disappearance some 100,000 missions in the skies of Vietnam.

CESSNA O-1F BIRD DOG

USAF

U.S. AIR FORCE   72814

Cessna O-1B/F Bird Dog

Few aircraft were as important for the efficient conduct of war operations in Vietnam as the small, unarmed Cessna O-1B, previously known as the L-19. Spearhead of the FAC (Forward Air Control), it formed part of the US Army organization until 1965, when all fixed-wing observation aircraft were turned over to the USAF. Flying at low level and reduced speed, their duty was to discover objectives, for the most part concealed in the jungle, such as groups of guerillas, convoys traveling along the Ho Chi Min Trail, or enemy units lying in ambush for unsuspecting government troops. Having spotted the enemy, they would immediately radio the DASC (Direct Air Support Center) which, as a rule, would be able to get attack aircraft to the spot within half an hour. The latter were again guided by the Bird Dog pilots who, in addition to pinpointing the objective with smoke or magnesium flares, would check the effectiveness of the strikes, if necessary correcting the aim. However, the O-1s were an easy target for the enemy, who could often hit them with ordinary rifle fire, without recourse to heavy anti-aircraft fire. Many Bird Dog pilots lost their life while carrying out their duty; they were usually officers with years of experience, veterans of many battles. Among the finest fighters, they succeeded in converting their little unarmed planes into formidable offensive weapons.

Martin B-57B/G Canberra

The first USAF jet warplanes to reach Vietnam after the Gulf of Tonkin crisis in August 1964 were 20 B-57 Canberras from the 13th and 8th TBS, the latter being the oldest in the USAF, formed as far back as 1917. They were preceded, in May 1963, by two RB-57Es, with electronic reconnaissance facilities which had already proved their immense worth. The Canberras began their round of missions on February 19, 1965 and were used until 1972, during the first few years chiefly for day and night attacks on the Ho Chi Minh Trail, later against other targets, including those in North Vietnam. Many of the B-57s of the 13th Bomb Squadron had been converted into B-57G models, equipped with sensors and relatively advanced electronic devices; the B-57 could carry in its bomb bay 14,260 anti-personnel bombs and 12 flares, as well as in wing pylons four 750lb (340kg) bombs; equally deadly were its four 20mm cannons. The B-57s were joined in their offensive forays by Mk.20 Canberras, manufactured by English Electric, from the 2nd Squadron of the Royal Australian Air Force, sent to Vietnam as part of the contribution of SEATO in 1969. Some 63 B-57s were shot down by Viet Cong and North Vietnamese anti-aircraft fire, indicating the scope of their activities and the danger of their missions.

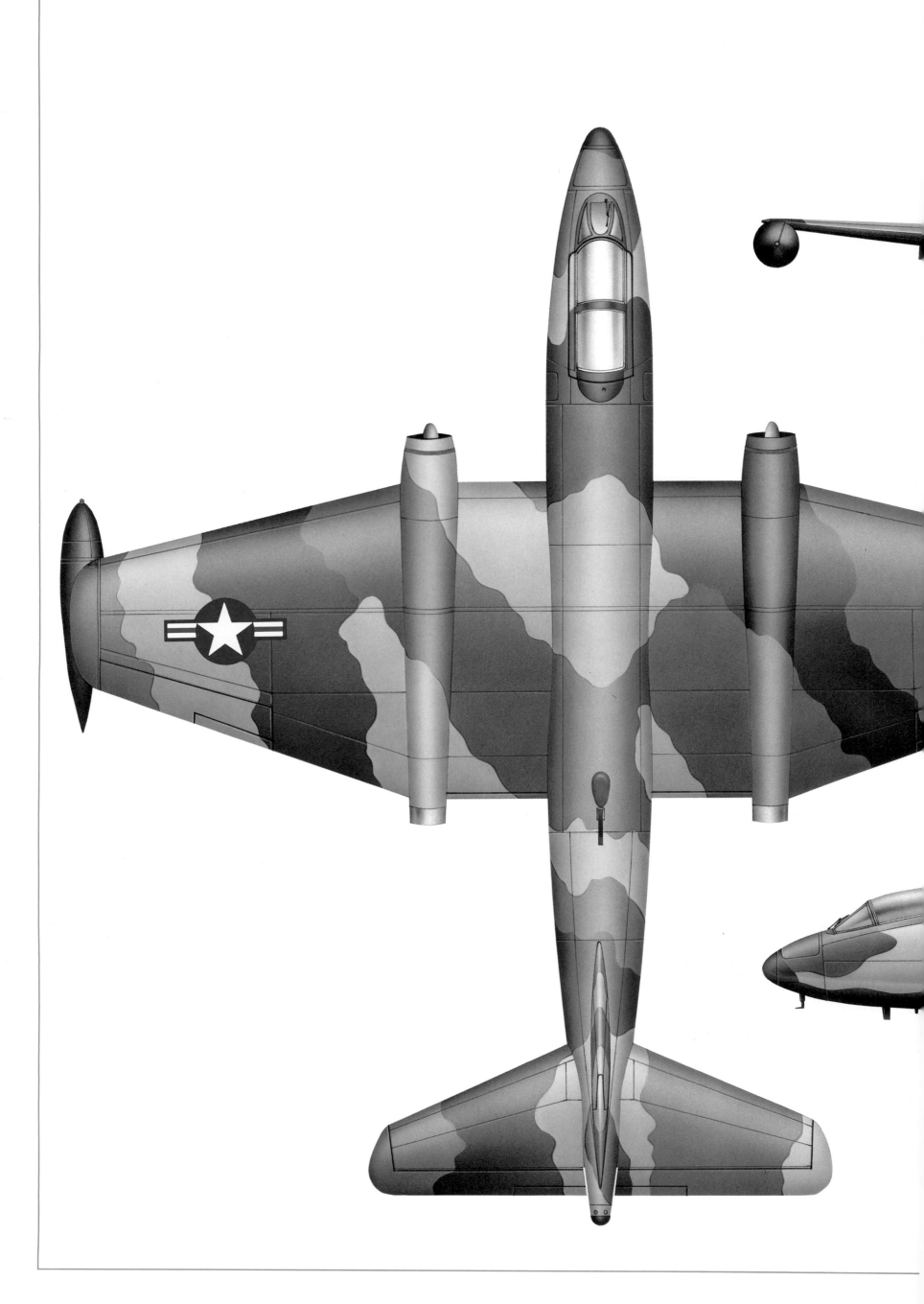

MARTIN B-57B CANBERRA

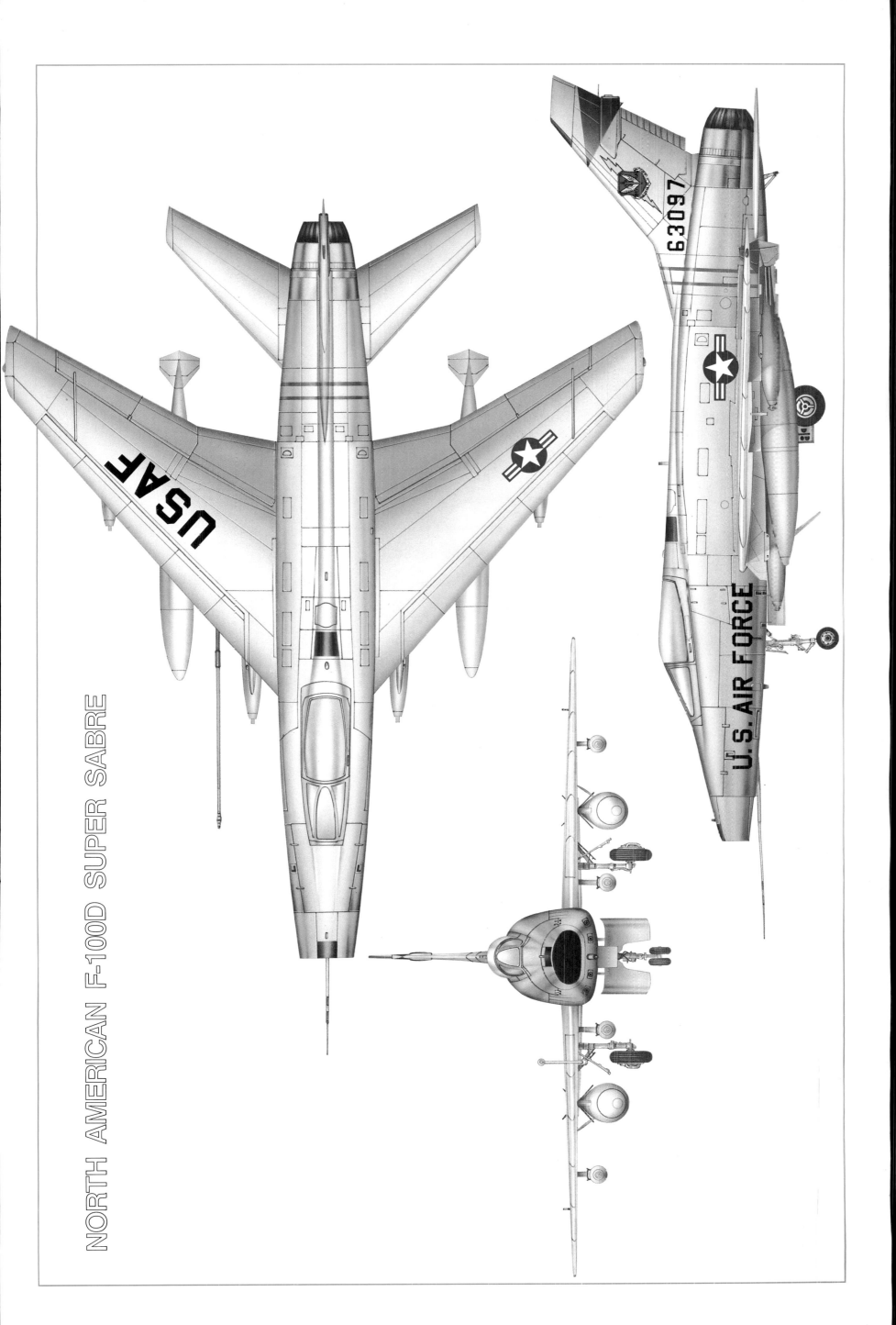

NORTH AMERICAN F-100D SUPER SABRE

North American F-100D/F Super Sabre

The F-100Ds, last single-seater version of the first American supersonic fighter, provided with an autopilot and also armed with bombs attached to the underside of the wings, played an important role in the Vietnam war, with over 300,000 missions from August 1964 to July 1971, when the 35th Tactical Fighter Wing finally left Phan Rang to return to the United States. The 615th TFS was the first unit furnished with F-100Ds to reach Da Nang on August 5, 1964, followed on August 17 by the 401st TFW, stationed at Tan Son Nhut. The Super Sabres, familiarly known as «Huns,» a shortened version of «Hundred,» were immediately used for low-level night bombing missions, and during the first years of the war pounded objectives in South Vietnam where suspected concentrations of Viet Cong had been sighted. For this type of mission the F-100Ds were armed with two CBU-24 bombs which, on opening, released a large number of anti-personnel devices, and two 750lb (340kg) napalm bombs. Once they had dropped their load, the Super Sabres proceeded to spray the zone under attack with their four 20mm cannons to complete the «cleaning up» work. Because of their adaptability and, even more, the lack of a real alternative, numerous F-100 Wings were used in Vietnam, some of them consisting of squadrons of the Air National Guard, called up for front line service. The 3rd TFW alone carried out more than 100,000 missions in 1969! From the end of 1965 a number of two-seater F-100Fs, the Wild Weasel I, carrying anti-SAM electronic equipment, were in action, operating from the Korat base in Thailand.

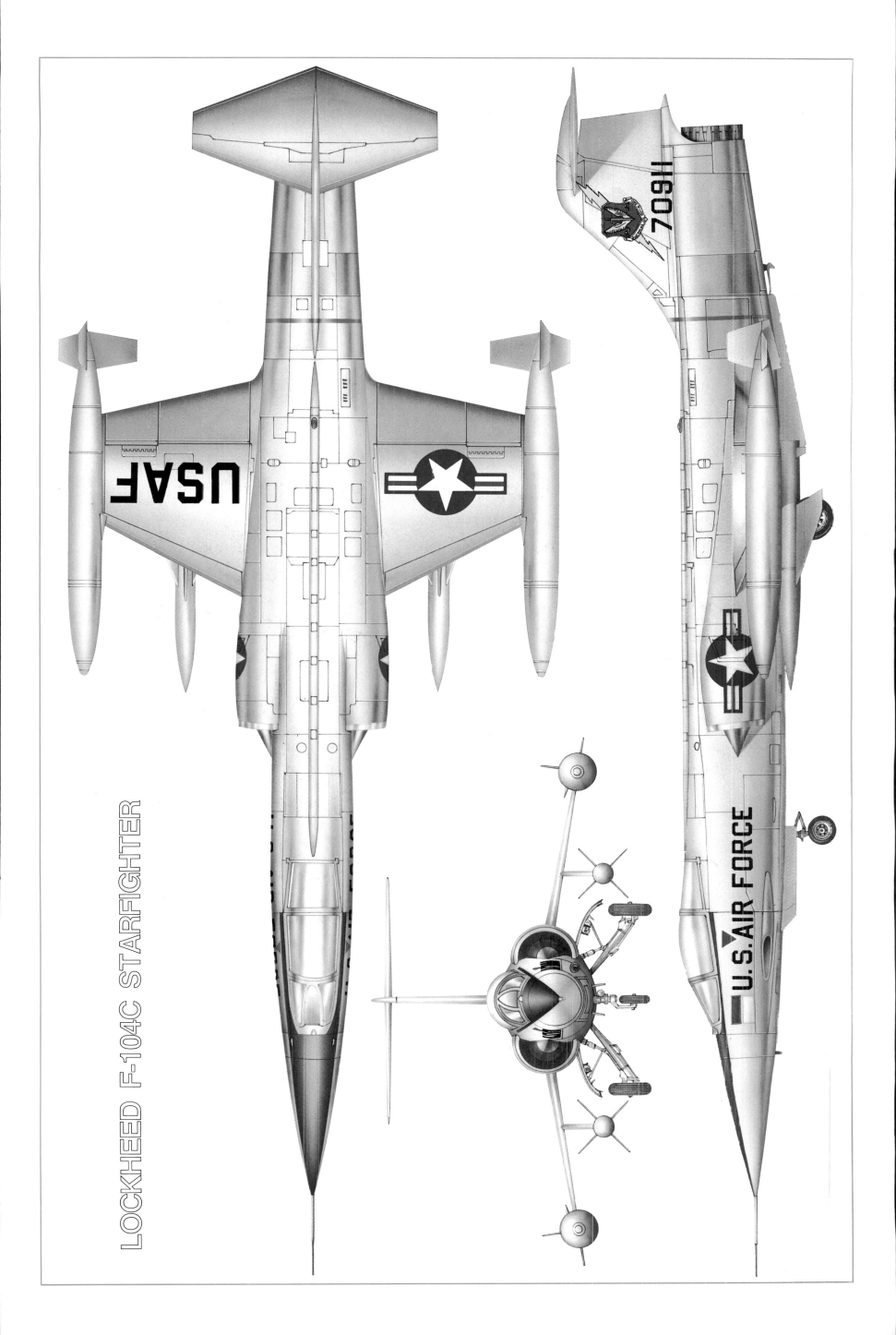

LOCKHEED F-104C STARFIGHTER

Lockheed F-104C Starfighter

Almost all the F-104s in the fighter-bomber version, assigned the letter C, were used in Vietnam, for 21 months, during which time they performed important, far-ranging work. The first fifteen Starfighters arrived in April 1965, with the 476th Tactical Fighter Squadron and the 479th Tactical Fighter Wing: from April 20 to November 20 of the same year they carried out 2,927 missions of machine-gunning, bombing and escorting strike aircraft, sometimes in North Vietnamese air space, before returning to the United States. The 476th was back, however, in June 1966, operating from the Udorn base in Thailand. The F-104Cs were now camouflaged in accordance with operational needs and in July were handed over to the 436th TFS and in October to the 435th, still belonging to the 479th TFW. From June 1966 to July 1967 the F-104Cs carried out escorting and bombing missions on North Vietnam, with over 5,290 sorties. The last Starfighter left Thailand before the end of 1967, its duties being taken over by the Phantom F-4D. Although not much has been written and said about the F-104C, it seems to have given a good account of itself, but the shortage of machines obviously limited its use.

Republic F-105D/G Thunderchief

The F-105 Thunderchief, familiarly called «Thud» by its pilots, received its baptism of fire in Vietnam and is indissolubly associated with that war, even though it was never used for the tactical atomic bombing for which it had been designed. In action in Vietnam from 1964 to 1970 was the single-seat F-105D, modified so as to carry bombs of the traditional type both in the hold and in underwing pylons, and also the two-seat F-105G, Wild Weasel, widely used for locating the radar emissions of SAM batteries, which they would either neutralize with their own electronic equipment, making enemy tracking impossible, or by means of direct bombing. All USAF squadrons furnished with the F-105 served in rotation in Southeast Asia, carrying out more than 20,000 offensive missions, and losing 330 planes, over a third of the total Thunderchief production. Employed without respite in strikes, mainly against North Vietnamese territory, the Thunderchiefs paid a heavy toll at the hands of anti-aircraft batteries, SAM missiles and enemy fighters, for they were easy prey unescorted and with a full bombload. It was rare for an F-105 pilot to complete his rota of 100 missions without being shot down at least once. On October 5, 1965, for example, in the attack on the Lang Met bridge, out of 24 F-105Ds of the 562nd Squadron of the 23rd TFW, only eight found their way back to their departure base in Thailand. Yet with their bombs they destroyed the objective. Free of their bombs, on the other hand, the F-105s were no sitting ducks for enemy fighters; from 1966 to 1967 they shot down 26 MiG-17s and one MiG-21 in air duels, a tally second only to that of the Phantom F-4s. The first F-105Ds arrived at Korat, n Thailand, in August 1964 with the 36th Squadron of the 7441st TFW, followed by those of the 18th, 355th and 388th TFW. To assess the importance of the Thunderchiefs as a strike force during the early part of the war, it is enough to point out that during 1965 threequarters of all attack missions against North Vietnam were carried out by this fighter-bomber, sometimes guided to its target by the Douglas EB-66 and subsequently escorted by Phantoms when the latter were thrown into action.

REPUBLIC F-105D THUNDERCHIEF

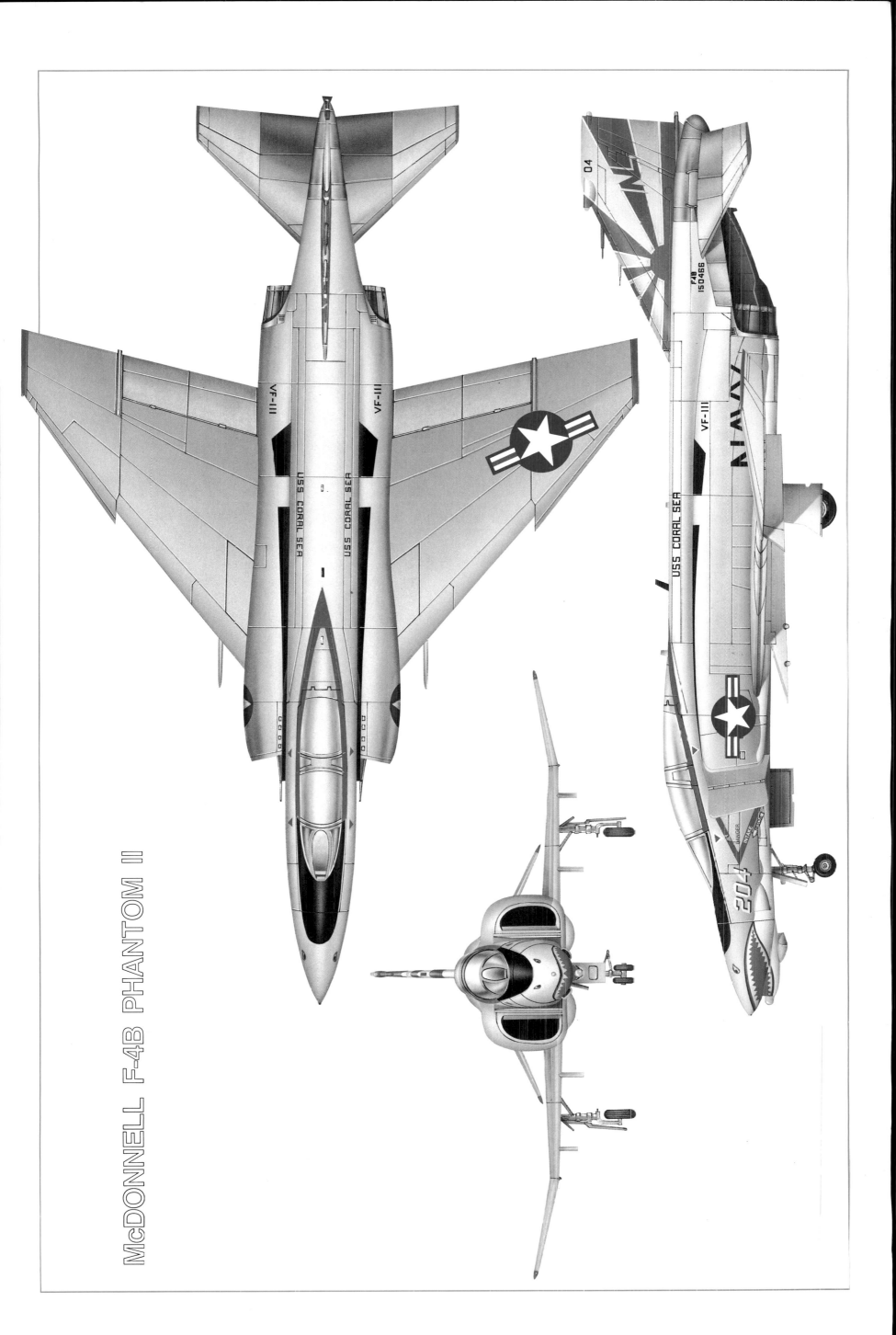

McDONNELL F-4B PHANTOM II

McDonnell F-4B/J Phantom II

Unquestionably the leading role in the air war in Vietnam was played by the McDonnell F-4 Phantom, which was used by the US Navy and the US Marine Corps in the B and J versions, and by the USAF in different versions. The earliest arrivals in the Southeast Asia theater of operations were the F-4Bs of the VMFA-531, on April 11, 1965. They were followed by those of many other Marine squadrons, based on land, and by squadrons of the US Navy operating from aircraft carriers. It would be unfair to single out the exploits of any particular unit because such a list would fill the pages of a sizeable book and because all distinguished themselves both in the attack/bombing role and in their demonstration of aerial supremacy. During direct encounters with the enemy, F-4Bs and F-4Js shot down 55 MiGs, of which eighteen were MiG-21s, two MiG-19s and the rest MiG-17s. Even so, it is fitting to mention the name of the US Navy Commander Randall H. «Duke» Cunningham, and of his radarman, Lieutenant William P. «Willie» Driscoll, with one MiG-21 and four MiG-17s to their credit, and to point out that the squadron boasting the biggest number of enemy planes downed was the VF-96, with eight certain victims and two probables. From the moment they went into action until the last day of the war, the Navy and Marine Phantoms never let up, gaining a reputation that they were later to emulate in other parts of the world.

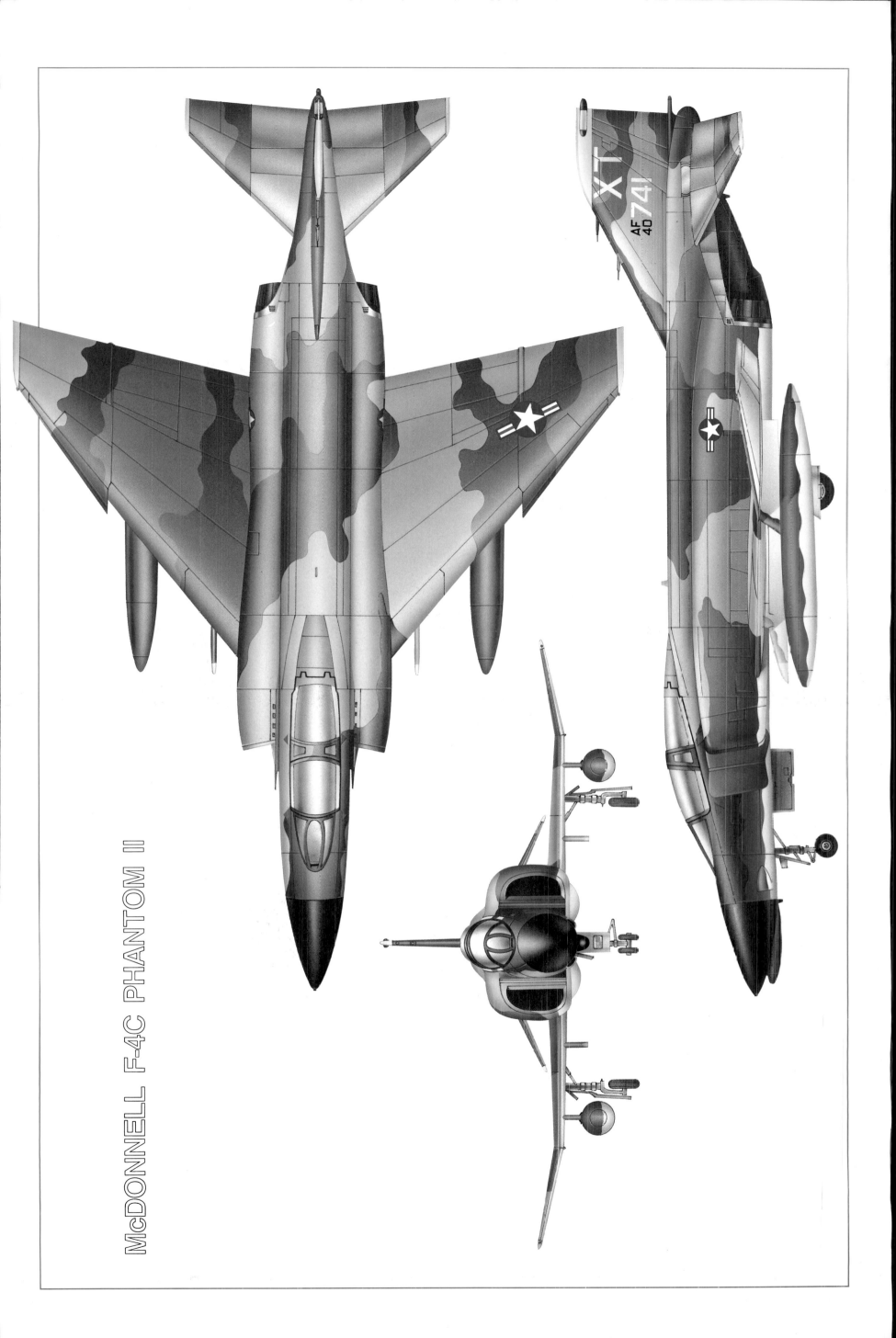

McDONNELL F-4C PHANTOM II

McDonnell F-4C/D/E Phantom II

The first F-4C Phantoms of the USAF arrived at the same time as those of the Marines, in April 1965, with the 45th Tactical Fighter Squadron, followed by those of the 12th TFW, in November, and of the 8th TFW, which was stationed at the Ubon base in Thailand in December of the same year. The Phantoms were detailed to play a defensive fighter role, escorting the F-105s weighted down with their bombloads, but when the ranks of the latter began to thin out, the Phantoms also took on attacking roles, achieving excellent results with their precision bombing. Although many units covered themselves with glory in eight years of war, it was the 8th TFW, among the first to reach Vietnam, which was most highly distinguished in battle. On January 2, 1967, F-4Cs of this Wing played a key role in the biggest aerial encounter of the war, shooting down seven MiG-21s without loss; and the 555th TFS (which with the 432nd and 433rd formed the 8th TFW) achieved more victories than any other USAF squadron, with a tally of 39 MiGs. In such dogfights the radar-controlled Sparrow air-to-air missiles and heat-sensitive Sidewinders proved invaluable, but the lack of a traditional cannon, notably for strike missions, soon became evident. It was for this reason that the F-4E version, sent into action toward the end of the war, was equipped with a rotary 20mm cannon, soon proving its worth not only in hitting the enemy on the ground but also in shooting down six enemy jets. By the end of the war the F-4s of the USAF boasted a record of 82 victories in air duels with MiGs, the success ratio in favor of the Phantom pilots being more than two to one.

Boeing B-52D/F Stratofortress

The only combat airplanes of the SAC (Strategic Air Command) to participate in the Vietnam war were the gigantic B-52s, tested in 1964 to guarantee the US the possibility of delivering nuclear strikes in virtually any part of the world. On February 17, 1965, B-52s of two squadrons based at Andersen, on the island of Guam, received orders to go into action in Vietnam. Each operation of proceeding to and returning from the objective entailed a journey of almost 5,500 miles (8,850km), with the need to refuel in the air and remain at least 12 hours in action. Despite these difficulties the B-52s were used to the full until 1971 to play a tactical role and subsequently a strategic function for which they were far better suited. Prior to April 1966 the Guam Strato-fortresses were of the B-52D type, modified so as to carry 750lb (340kg) bombs, to a total weight of 42,075lbs (19,000kg) both in the main bay and under the wings. After April the B-52s which went into action were radically transformed, with a capability of carrying in the main bay 66 bombs weighing 750lb (340kg) or 85 bombs each of 500lb (227kg) and another 24 under the wings, giving a total payload of 74,250 lbs (33,680kg). During the early years of the war, the «Buffs» or «Beasts,» as the B-52s were familiarly called (Big, Ugly, Fat Fellow being other variations), were used chiefly in South Vietnam and on the Ho Chi Minh Trail, in the so-called Arc Light operation. In January 1968 they achieved miracles in the attempt to relieve the pressure of the North Vietnamese siege of the entrenched Khe Sanh camp, defended by the Marines. The planes succeeded in dropping their deadly load at a mere 2,624 ft (800 m) from the base perimeter, causing substantial enemy losses. During the siege the B-52s dropped 54,000 tons of bombs on the objective in the course of 2,548 missions. B-52s played an equally significant role during the strategic bombardment of North Vietnam: in June 1972 alone more than 200 Stratofortresses from the Guam base carried out no less than 100 missions daily, not counting those accomplished by the other B-52 units stationed from April 1967 at the U-Tapao base in Thailand, only two hours flight from the target. In October 1972 President Nixon ordered the resumption of heavy bombing of North Vietnam, and the Guam B-52s made 729 sorties to drop 15,000 tons of bombs on the enemy between December 18 and 29. During eight years of operations the B-52s of SAC carried out some 7,784 missions, dropping 2,765,000 tons of munitions on the enemy. Although it is difficult to estimate the material damage caused by the Stratofortresses, their psychological effect on the enemy was certainly immense. Quite a number of B-52s were destroyed in action, mainly by SA-2 ground-to-air missiles, so that it is hard to decide whether the means really justified the ends; what cannot be disputed is that the crews of the SAC must rank with the many serv-icemen in Vietnam who performed up to and beyond the bound of duty.

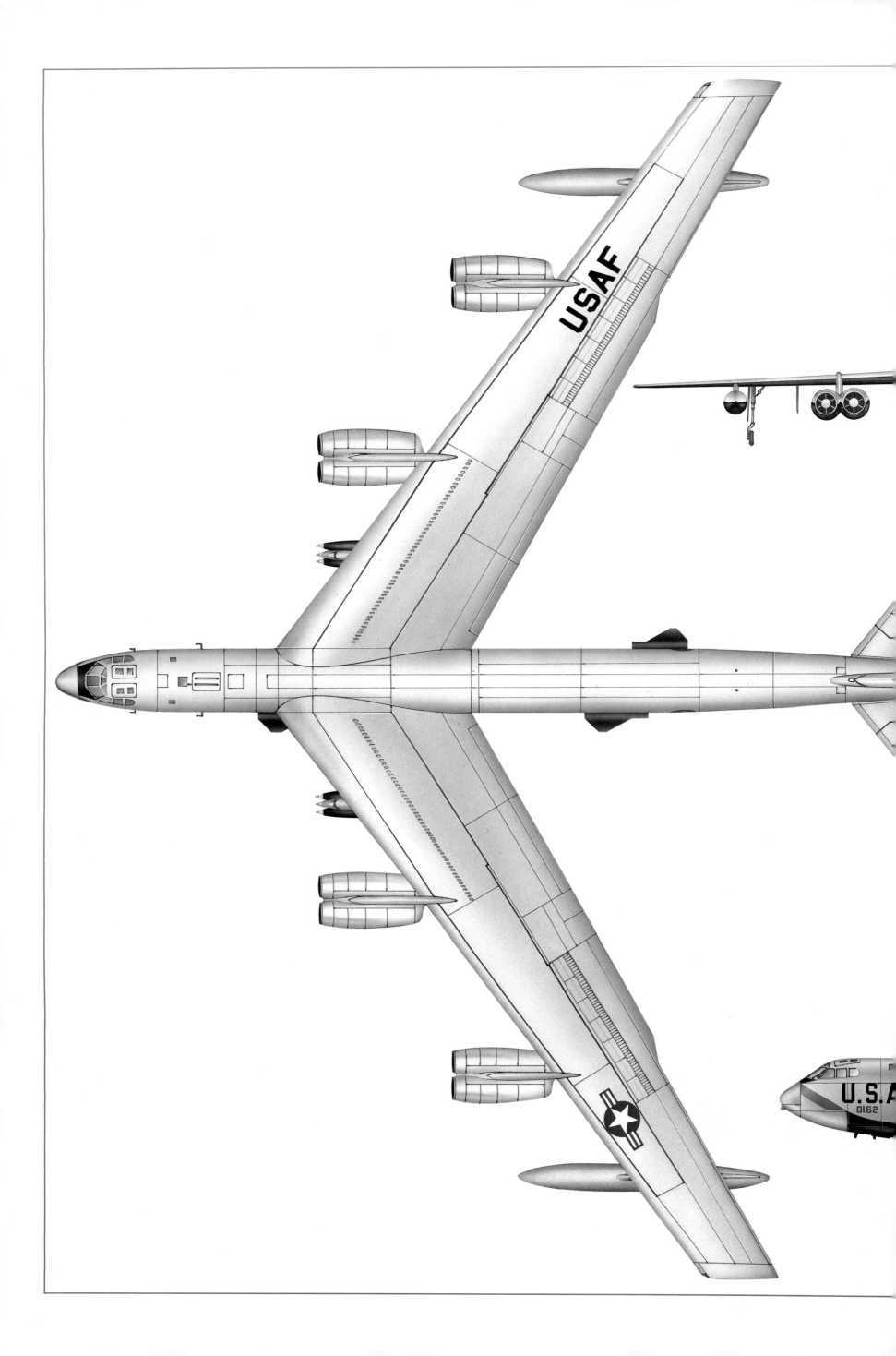

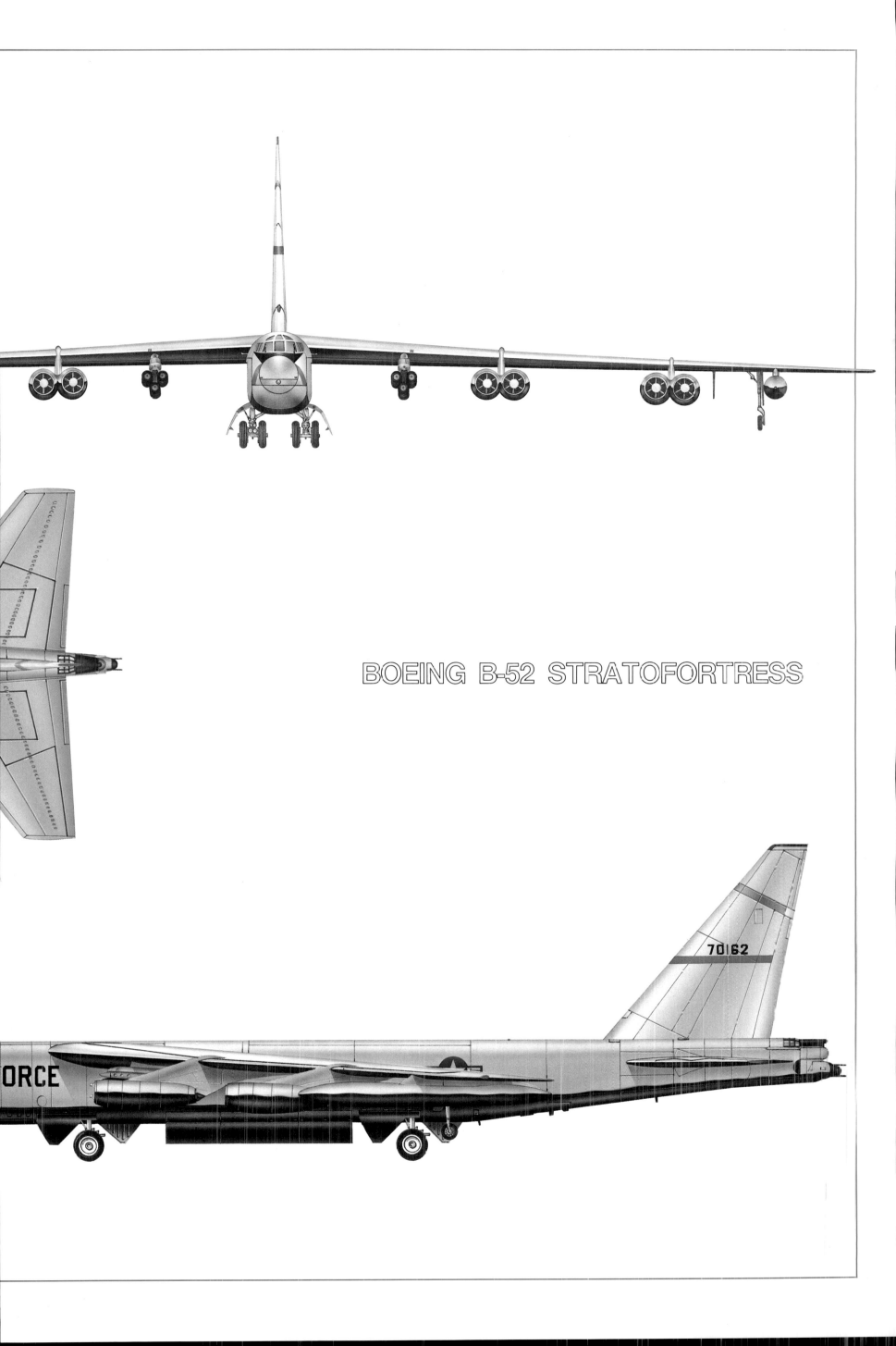

BOEING B-52 STRATOFORTRESS

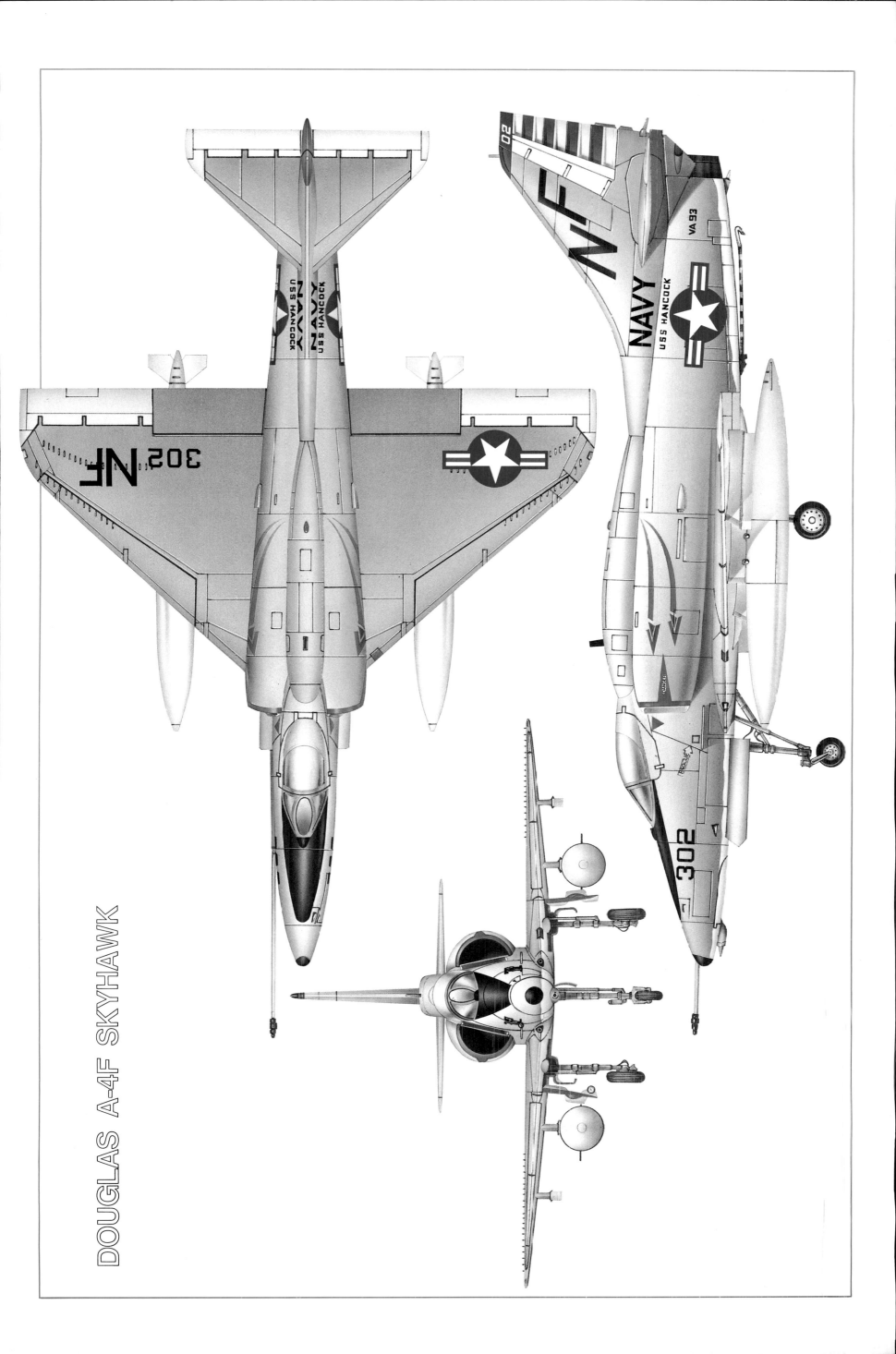

DOUGLAS A-4F SKYHAWK

Douglas A-4F Skyhawk

Ideal successor to the marvelous Skyraider, the Douglas A-4, a daytime fighter-bomber, was the favorite plane of land-based Marine units and was also widely used by US Navy squadrons, particularly after 1968. Some idea of the Skyhawk's capabilities can be judged by the fact that a single Marine squadron, the VMA-311, carried out a record number of 47,663 sorties between June 1965 and May 1971. The US Navy was so convinced of the aircraft's adaptability to this type of warfare that in 1966, after the initial experiences on the battlefield, it gave the order for production to be resumed so as to take delivery of a further 146 examples of the A-4F. Of all the planes used by the US Navy and US Marine Corps to carry out strike missions in Vietnam, the Skyhawks alone were responsible for over 50 percent; and they suffered the highest battle casualties, losing 196 machines, mainly because they were most often exposed to anti-aircraft fire and enemy fighters.

GRUMMAN   A-6A INTRUDER

GENERAL DYNAMICS F-111A

Vought F-8E/J Crusader

Central to the Tonkin Gulf incident, which led to America's direct intervention in the Southeast Asian crisis, was the Vought F-8E, which on August 2, 1964, defended US Navy units from attack by North Vietnamese motor torpedo boats. Four Crusaders of the VF-53 Squadron from the aircraft carrier USS Ticonderoga, sunk an enemy vessel with missiles and gunfire, thus initiating a war that did not end until 1973. From 1964 to 1969, during which time the Crusaders were gradually replaced on US Navy aircraft carriers by F-4B Phantoms, the F-8s, designed as daytime supersonic fighters, were also largely employed in strike missions mostly over North Vietnam, carrying up to 5,000lbs (2,268 kg) of bombs under the wings. The first encounter between F-8Es and MiG-17s took place on July 12, 1967, and after that date the Crusaders frequently took on the North Vietnamese fighters, emerging from battle with a tally of fifteen MiG-17s and three MiG-21s shot down in dogfights, for the loss of only three planes. However, another 53 F-8Es and F-8Js fell victim to North Vietnamese anti-aircraft batteries, and a further 58 were destroyed while in action as a result of various causes. An important support role was also played by the RF-8 reconnaissance planes, 38 of them being lost through anti-aircraft fire, SAM missiles or accidents. All US Navy aircraft carriers engaged in the war were equipped with Crusader squadrons, the most successful of these being the VF-121 which chalked up six victories in fights with MiGs. Altough there was no cause for regret in their replacement by the McDonnell F-4B Phantoms, the Crusaders ranked third as «MiG killers» in Vietnam, after the Phantoms themselves and the Republic F-105s.

VOUGHT A-7D CORSAIR II

DOUGLAS AC-47B GUNSHIP

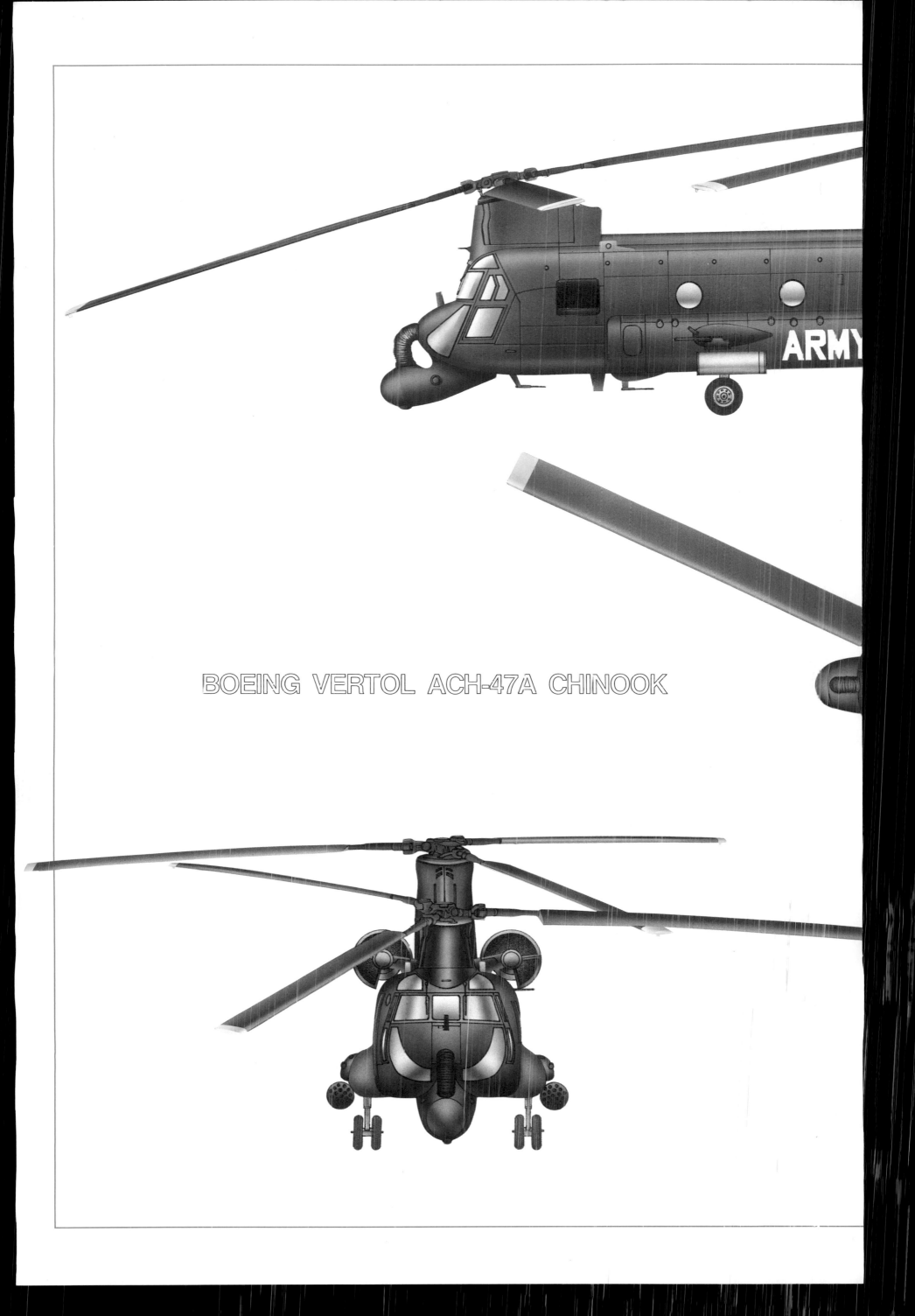

BOEING VERTOL ACH-47A CHINOOK

Boeing Vertol ACH-47A Chinook

Elder brother of the omnipresent Huey and equally valuable for a wide variety of activities involving men and materials, from landing and reinforcement to rescue and evacuation, but chiefly for transport in the battle zone, the Chinook also boasted a combat version, the ACH-47A. In June 1965 the Army Material Command ordered from Boeing four CH-47s modified as «Gunships,» armed with cannon, machine guns and rockets, armor-plated, and with bigger, self-sealing fuel tanks for longer flight capacity. On May 13, 1966, three of the four ACH-47As arrived at the Vung Tau base, consigned to the 53rd Aviation Detachment as part of the 147th Aviation Company, one of the first units to receive CH-47s in Vietnam. Joined several months later by the fourth machine, they came through a long testing period with much credit, even though the US Army did not consider it opportune to order their mass-production. The four ACH-47As, nevertheless, continued to operate very usefully on the battlefield until, one by one, they were destroyed. Of the four, one was salvaged and is now in the United States. It has to be remembered that the Chinook, in the course of its normal duties, proved an exceptional machine, rescuing some 11,000 helicopters and planes that had been immobilized either by crashing accidentally or being hit by enemy fire in landing operations. An incredible tally, yet true: in one evacuation mission a Chinook managed to take aboard 147 men, almost four times the normal load envisaged by its manufacturers, namely 44 soldiers in combat gear.

Rockwell OV-10A Bronco

The only real COIN (Counter Insurgency) plane to take part in the Southeast Asian war was the Rockwell OV-10A Bronco. The first of these were sent with the Marines to the operation zone as soon as they came off the assembly line. Subsequently used both by the US Navy and the USAF, this tactical reconnaissance plane proved extremely useful and well suited to the requirements of the FAC (Forward Air Control). Very often, confident of its own fire power, the Bronco would strike a target without even waiting for other warplanes to arrive. It went into action in 1968, barely two years after the first flight of the prototype, but did not have the chance to be used in such numbers as other planes which were admittedly less suited for the difficult and dangerous job of being the advance «eye» of the DASC (Direct Air Support Center).

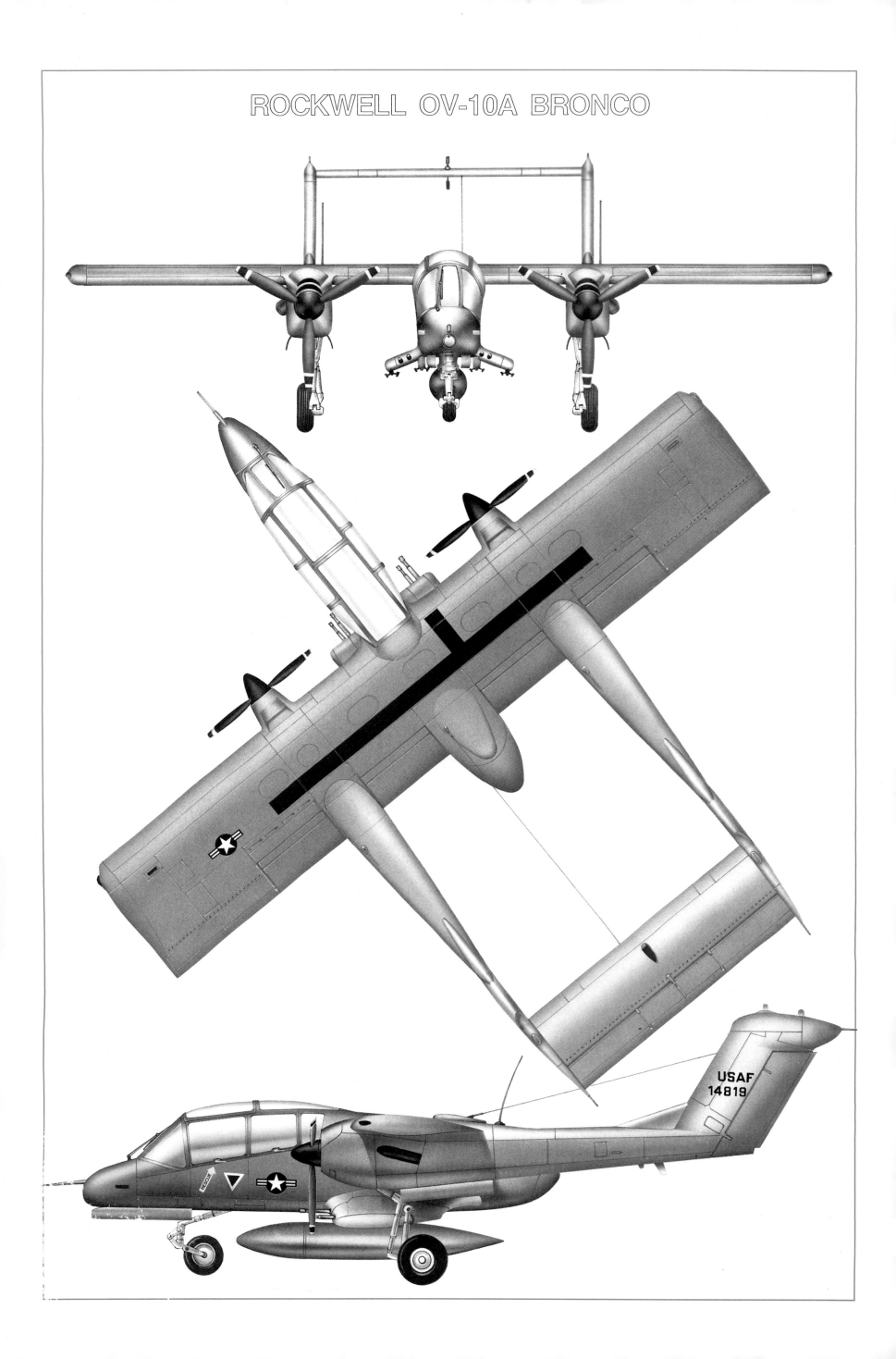

ROCKWELL OV-10A BRONCO

Northrop F-5A/E Freedom Fighter

In August 1964 the USAF took its first delivery of F-5As and immediately decided to send a few machines to the combat zone in order to test their capabilities. The so-called Skoshi Tiger program was organized in October 1965, coinciding with the arrival of 12 F-5As, partially modified for war purposes and furnished with «proboscis» equipment for refueling in flight. The fighter-bombers operated at first with the 4503rd TFW, and in the course of 2,500 hours of tactical support and reconnaissance missions gained experience that proved extremely valuable for launching the next, more powerful F-5E version, which took the name Tiger II in recognition of the aircraft's contribution to the Skoshi Tiger operation. The 12 F-5As of the 4503rd TFW, together with six new machines, were handed over to the 10th Fighter Command Squadron, attached to the 3rd TFW at Bien Hoa, and in 1967 the USAF delivered them to the VNAF. The F-5s were the first and only jet aircraft belonging to the newly formed South Vietnamese air force, which later received a number of F-5Es, used in action until the final collapse. Many of these F-5Es were captured by the North Vietnamese in perfect working order.

NORTHROP F-5A FREEDOM FIGHTER

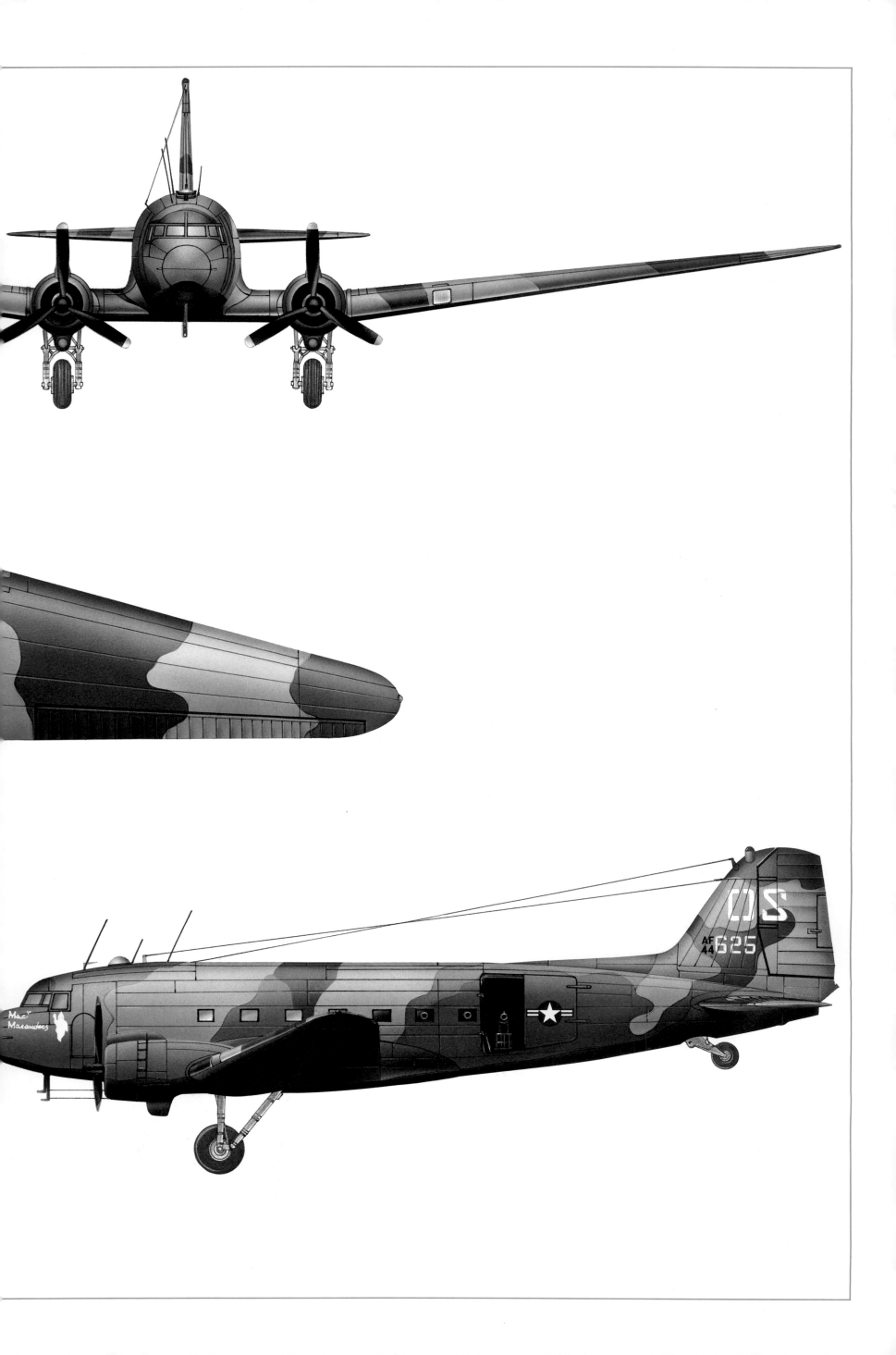

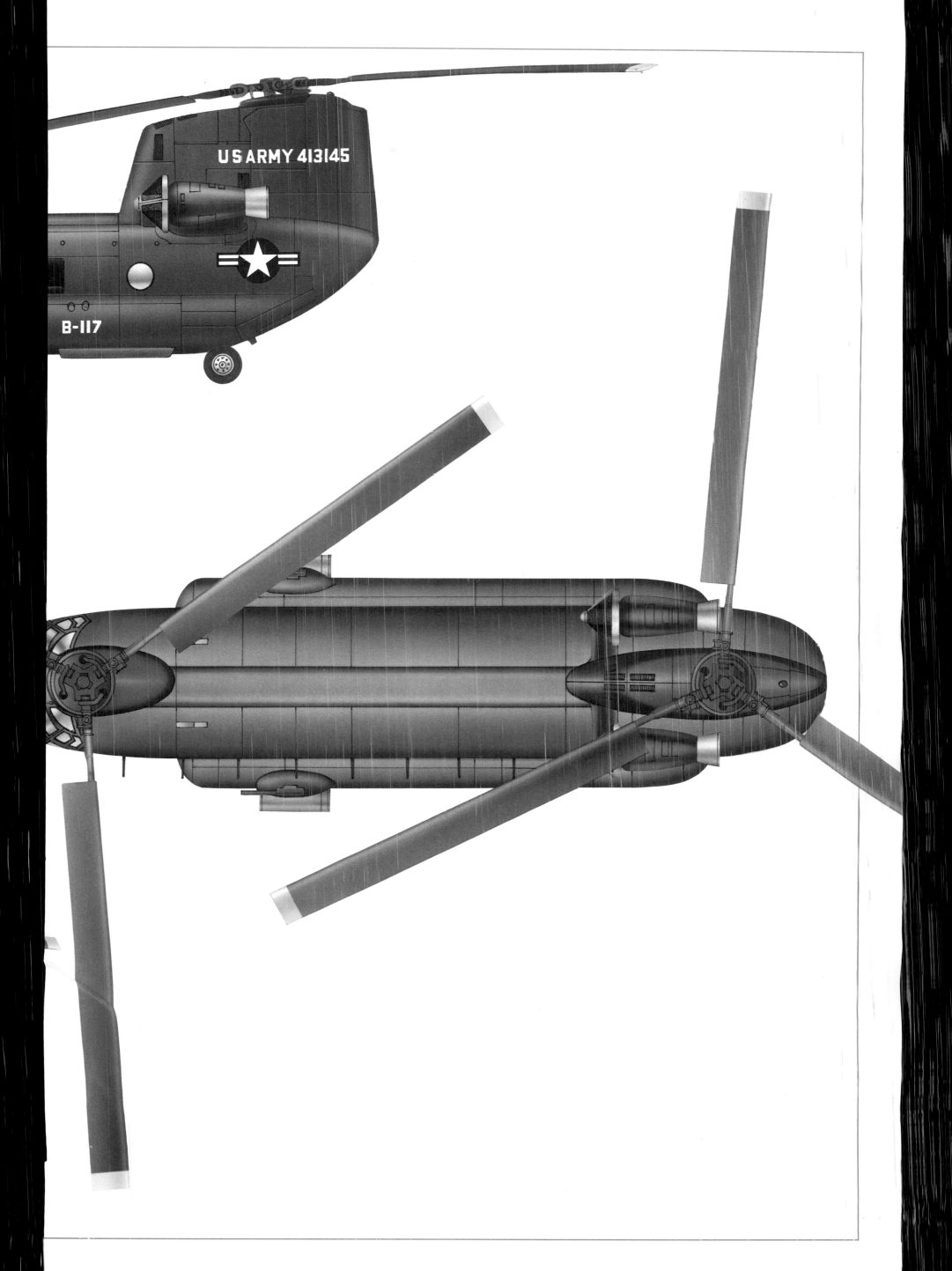

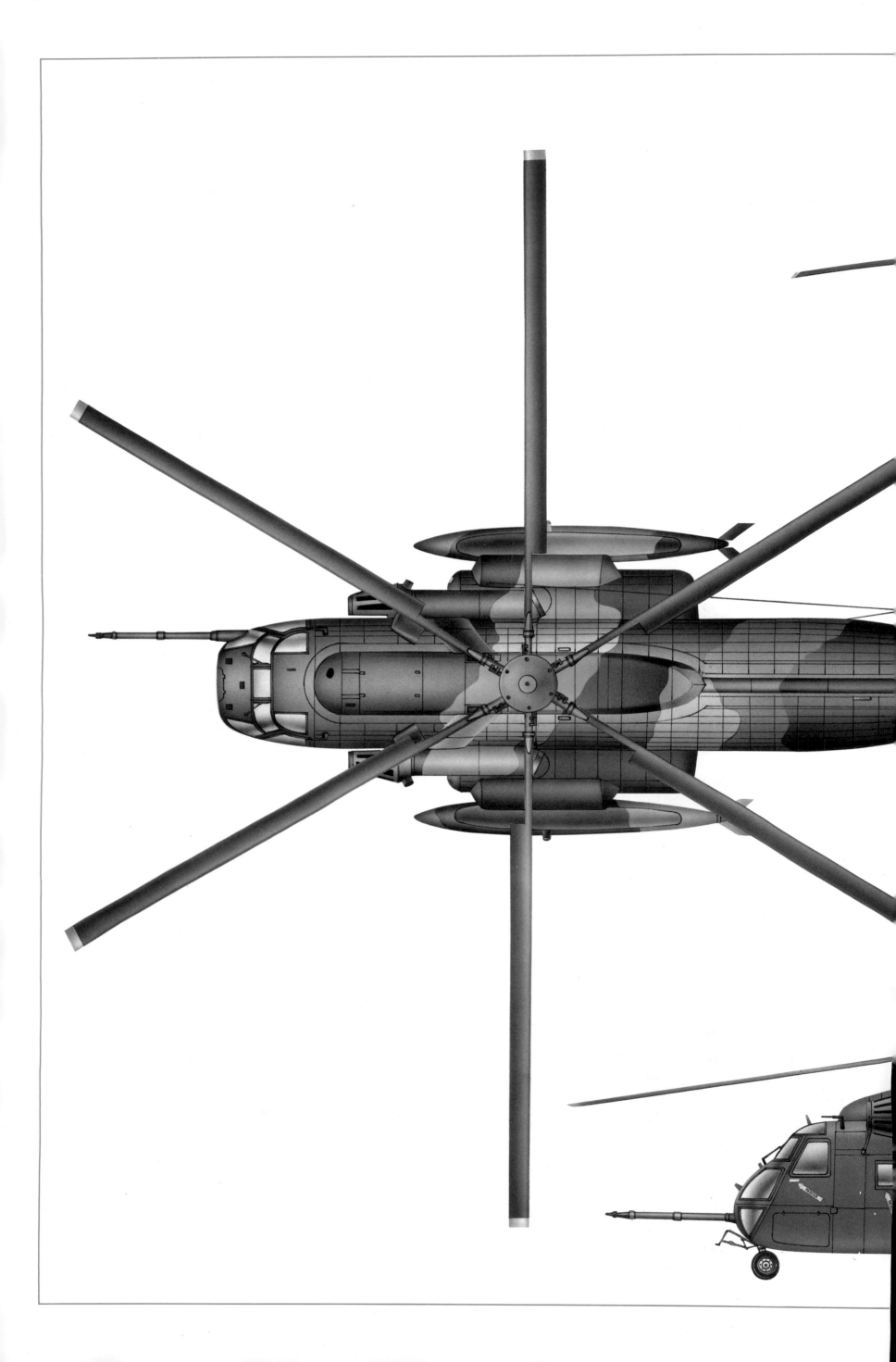

SIKORSKY HH-53C SUPER JOLLY

USAF
5784

Cessna A-37B Dragonfly

One of the few aircraft designed from the start for tactical support arrived in Vietnam toward the end of the 1960s and was mainly used in support of helicopter operations. Capable of mounting a wide range of weapons, it proved highly adaptable to diverse operational needs. Particularly effective were its low-level napalm bomb attacks. A fairly limited number of machines, under the colors of both the USAF and the VNAF (the illustration shows the VNAF insignia), were used in action.

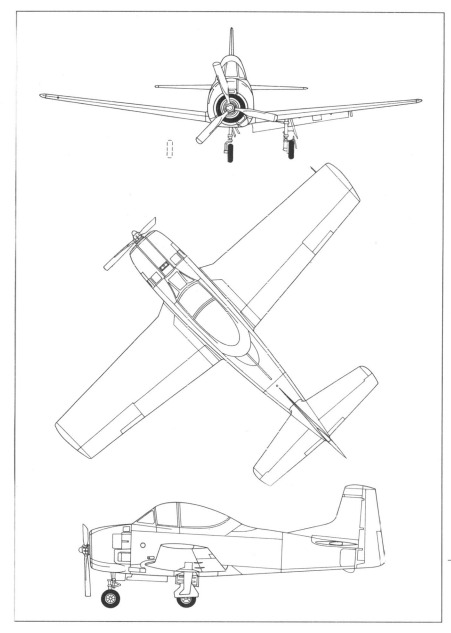

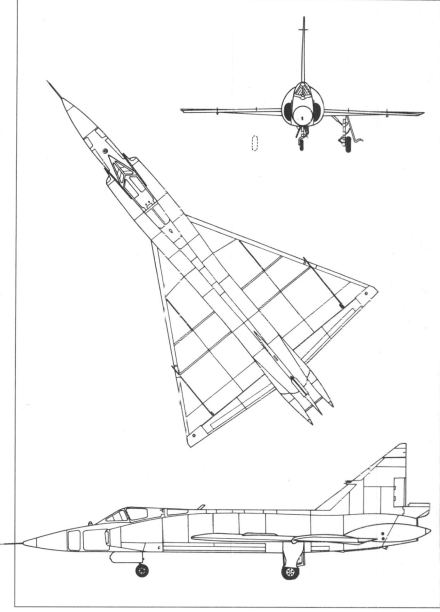

North American T-28

The North American T-28 was the last propeller trainer made in the United States. Known as the Trojan, it was designed in 1948 to replace the by-now ancient T-6 Texan. The first prototype flew on September 26, 1949, and deliveries of the T-28A version (destined for the USAF) began the following year. Up to 1956 in all, 1,194 machines were used in front line units, and up to 1959 in reserve units. In 1952 the US Navy also ordered the T-28 in the B version (more powerful engine, 489 planes) and the C version (299 planes). These aircraft went into service in 1953 and continued until the late 1960s. The T-28, however, was not only used for training: it was transformed in 1962 into a ground attack and anti-guerilla plane, the T-28D version. Many of these aircraft were used in Vietnam, but most of them were taken over by the French Armée de l'Air which, by 1950, had bought 245 of them — when withdrawn from the USAF — putting them into service under the name of Fennec.

Aircraft: North American T-28D
Year: 1962
Type: trainer-attack
Manufacturer: North American Aviation
Engine: Wright R-1820-86 9 cyl. radial, air cooled
Power: 1445hp
Wingspan: 40ft 7in (12.37m)
Length: 32ft 9in (9.98m)
Height: 12ft 7in (3.84m)
Wing area: 268sq ft (24.90m²)
Max take-off weight: 8,250lb (3,742kg)
Empty weight: 6,420lb (2,912kg)
Max speed: 352mph at 18,000ft (566km/h at 5,490m)
Service ceiling: 37,000ft (11,280m)
Range: 1,200mi (1,931km)
Crew: 2
Load-armament: pods with GE minigun, 500lb (bombs, rockets on six underwing pylons)

Convair F-102

The first delta-wing interceptor to be used by the USAF, the Delta Dagger was also the first supersonic plane to be fitted exclusively with air-to-air missiles. First planned in 1951, Convair, in developing it, made full use of the study and research which had culminated three years previously with the production of the experimental XF-92A. The first YF-102 prototype flew on October 24, 1953, but succeeding tests brought to light a series of complex aerodynamic problems, which were resolved only a year later, with the appearance (first flight December 20, 1954) of the YF-102A. Production began immediately and deliveries to the USAF started in June 1955, ending in April 1958 with the last of the 875 planes ordered (plus 63 TF-102A two-seater trainers). In service from April 1956, the Delta Daggers were used in Vietnam for six years. The F-102As remained in service in units of the Air National Guard until 1978.

Aircraft: Convair F-102A
Year: 1955
Type: interceptor
Manufacturer: Convair Division of General Dynamics
Engine: Pratt & Whitney J57-P-23
Power: 16,000lb (7,257kg)
Wingspan: 38ft 1¹/₂in (11.60m)
Length: 68ft 4¹/₂in (20.82m)
Height: 21ft 2¹/₂in (6.45m)
Wing area: 695sq ft (64.56m²)
Max take-off weight: 31,500lb (14,288kg)
Empty weight: 19,350lb (8,777kg)
Max speed: 825mph at 36,000ft (1,328km/h at 10,973m)
Service ceiling: 53,400ft (16,276m)
Range: 386mi (621km)
Crew: 1
Load-armament: 6 missiles; 24 rockets

North American T-28D over South Vietnam in 1962.

Convair F-102A of the 509th Fighter Interceptor Squadron.

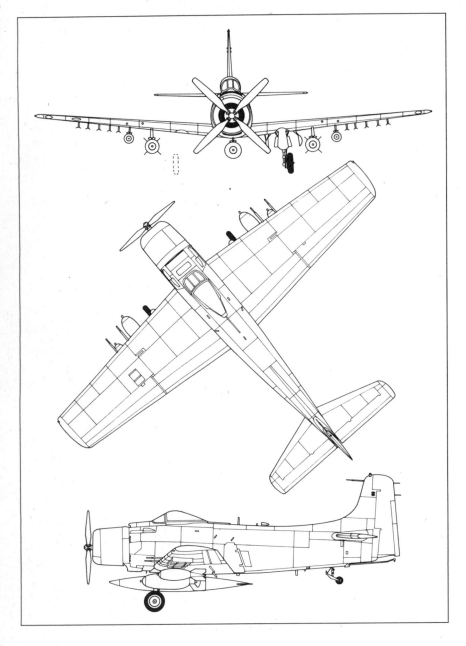

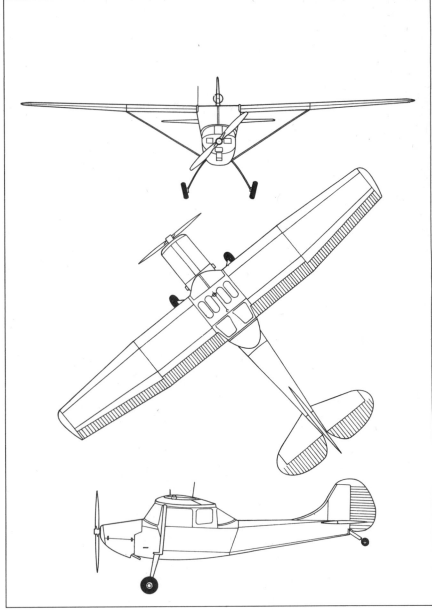

Douglas A-1

The Skyraider was the last great single-seater combat plane with piston engine. Designed toward the end of World War II as a divebomber, and torpedo-carrier, it proved itself more versatile and efficient than even its makers had imagined. The Douglas AD (as it was designated) proved its worth during the jet era, first in Korea and then in Vietnam, so much so that in 1966 (nine years after the closure of the production lines which, from 1947 to 1957, had turned out 3,180 aircraft of seven basic versions) serious consideration was given to resuming production. The first prototype flew on March 18, 1945 and operational service began in December 1946, with the AD-1 (277 machines). There then followed 178 AD-2s, 193 AD-3s and 1,051 AD-4s. All these versions had four main roles: daytime and all-weather attack, radar patrol, and electronic countermeasures. In 1951 the variant

Aircraft: Douglas A-1H
Year: 1952
Type: attack
Manufacturer: Douglas Aircraft Co.
Engine: Wright R-3350-26WA, radial, 18 cyl., air cooled
Power: 2738hp
Wingspan: 50ft 0in (15.24m)
Length: 39ft 2in (11.83m)
Height: 15ft 8in (4.77m)
Wing area: 400.33sq ft (37.192m²)
Max take-off weight: 25,000lb (11,340kg)
Empty weight: 11,968lb (5,429kg)
Max speed: 322mph at 18,000ft (518km/h at 5,846m)
Service ceiling: 28,510ft (8,690m)
Range: 1,142mi (1,840km)
Crew: 1
Load-armament: 4x20mm cannon; 7,960lb (3,630kg)

two-seater AD-5 appeared, with a bigger cabin, and a year later production resumed of the single-seater AD-6 (713 machines). The last series was the AD-7 (72 planes) in 1955.

Cessna O-1

The Cessna Bird Dog was the most popular light aircraft used by the US Army for liaison and observation in the postwar period. More than 3,500 machines left the assembly lines from the end of 1950 and remained in service until the late 1970s, taking part in the Korean War and Vietnamese War. The Bird Dog was derived directly from the Cessna 170, a commercial model in production in 1950. From the first order for fourteen planes in June 1950, the numbers increased dramatically, until by October 1954 the total production of L-19As (as they were originally designated) was 2,486. Two years later another 310 TL-19D training planes were ordered, while in 1957 the final version appeared, namely the improved and more powerful L-19E, which brought the total production to 3,431 machines. In 1962 the different versions were renamed, in sequence, O-1A, O-1B, TO-1D and O-1E.

Aircraft: Cessna O-1E
Year: 1956
Type: observation
Manufacturer: Cessna Aircraft Co.
Engine: Continental O-470-11, 6 cyl., air cooled
Power: 216hp
Wingspan: 36ft 0in (10.97m)
Length: 25ft 10in (7.87m)
Height: 7ft 4in (2.23m)
Wing area: 174sq ft (16.16m²)
Max take-off weight: 2,400lb (1.090kg)
Empty weight: 1,614lb (732 kg)
Max speed: 130mph (209km/h)
Service ceiling: 18,500ft (5,640m)
Range: 530mi (853km)
Crew: 3

Douglas A-1H of the 1st Special Operations Squadron.

Cessna O-1G of the 19th Tactical Air Support Squadron.

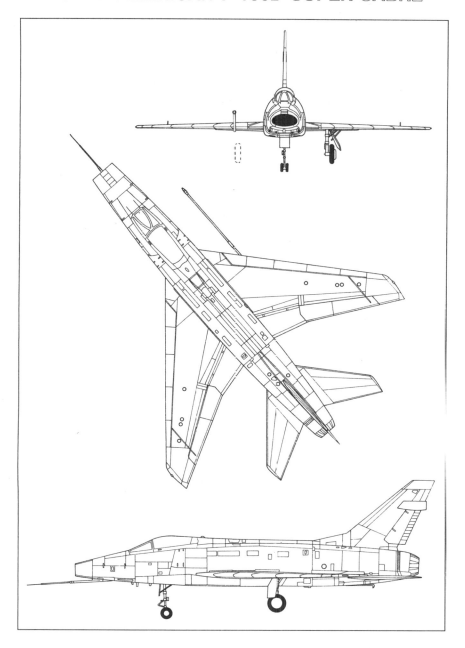

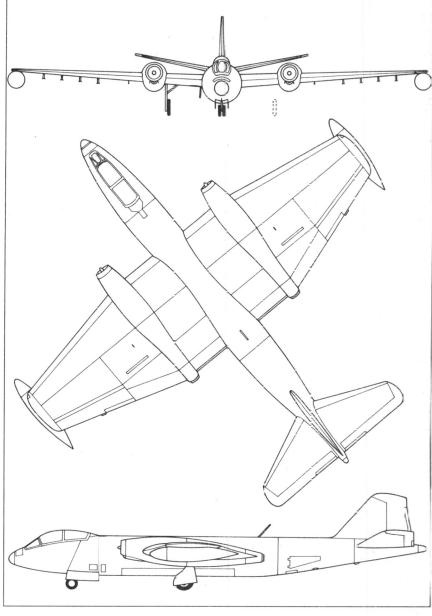

North American F-100

First supersonic fighter in the West, the F-100 originated in the early 1950s as direct successor to the famous F-86 Sabre. The USAF ordered two prototypes on November 1, 1951, and the first of these took off on May 25, 1953. Mass-production began immediately with the initial A variant (203 machines). In the second C version the role of interceptor was transformed into that of fighter-bomber: ordered in February 1954, 476 machines were built. Then followed (first flight January 24, 1956) the F-100D version, produced in largest numbers, with 1,274 machines. The last was the F-100F, a two-seater training plane (first flight March 7, 1957), of which 339 were built. The Super Sabre, in service from 1954, first saw action ten years later in Vietnam, where it remained in service until 1971. These planes were used principally for tactical support work.

Aircraft: North American F-100D
Year: 1956
Type: fighter-bomber
Manufacturer: North American Aviation
Engine: Pratt & Whitney J57-P-21A
Power: 16,950lb (7,688kg)
Wingspan: 38ft 9in (11.81m)
Length: 50ft 0in (15.24m)
Height: 16ft 2 ³/₄in (4.95m)
Wing area: 400sq ft (37.16m²)
Max take-off weight: 34,832lb (15,800kg)
Empty weight: 21,000lb (9,526kg)
Max speed: 864 mph at 36,000ft (1,390km/h at 10,973m)
Service ceiling: 36,100ft (11,003m)
Range: 534mi (859km)
Crew: 1
Load-armament: 4x20mm cannon; 2 missiles; 7,500lb (3,401kg)

Martin B-57

In 1951, after the failure of the projected Martin XB-51 bomber, the USAF decided to order the twin-engined English Electric Canberra. Although many were doubtful, the choise proved to be excellent: the plane responded perfectly to all handling demands, versatility, operational flexibility, range and bombload capacity. Known as the B-57, the aircraft was produced under license by Martin itself, which started production with eight examples of the pre-production B-57As (first flight July 20, 1953) and 67 RB-57A reconnaissance planes. The first B-57B (202 machines) appeared on June 28, 1954, and became operative the following year. Then followed 38 B-57C training planes and 68 multipurpose B-57Es. Martin also built 20 RB-57D reconnaissance planes, with longer wings to improve performance at high altitude. These aircraft, however, showed signs of fatigue and were grounded in 1963. As a result,

Aircraft: Martin B-57B
Year: 1954
Type: bomber
Manufacturer: Glenn L. Martin Co.
Engine: 2 x Pratt & Whitney J65-W-5
Power: 7,220lb (3,275kg)
Wingspan: 64ft 0in (19.50m)
Length: 65ft 5in (19,96m)
Height: 15ft 7in (4.75m)
Max take-off weight: 55,000lb (24,950kg)
Empty weight: 26,800lb (12,150kg)
Max speed: 582mph (937km/h)
Service ceiling: 48,000ft (14,630m)
Range: 2,100mi (3,380km)
Crew: 2
Load-armament: 8 machine guns; 6,000lb (2,720kg)

General Dynamics was given the job of designing another photoreconnaissance variant, the RB-57F, of which 21 machines were built. The last version was the B-57G, an up-to-date form of many early models, fitted out to the most modern requirements.

North American F-100D of the 429th Tactical Fighter Squadron.

Martin B-57 Canberra.

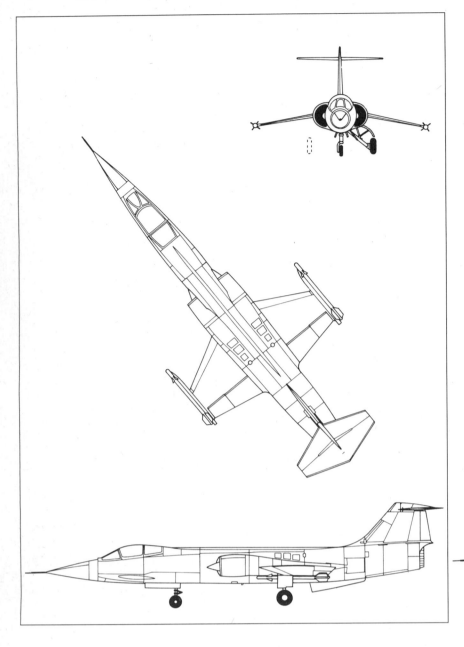

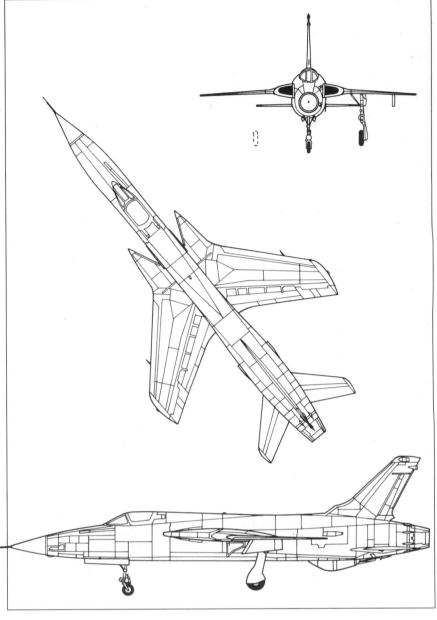

Lockheed F-104

First USAF fighter to fly above Mach 2, the F-104 Starfighter made its appearance in the 1950s when it was decided to replace the still airworthy F-100 with a fighter which could be used mainly as an interceptor. Planning started in 1952 and the first of two prototypes took to the air on March 4, 1954. Seven months later came the initial order for 153 machines of the F-104A series, followed by 26 two-seater F-104B trainers. Despite its exceptional qualities, however, the USAF considered it unsuitable for interception alone, and with the C version (77 machines, first delivery October 16, 1958), the F-104 was transformed into a fighter-bomber. This aircraft had a brief operational life in Vietnam. However, the Starfighter's fortunes were lifted by production of the next G version for the NATO allies. From 1960 to 1973 some 1,127 of this variant were produced under license in Canada, Japan, Belgium, Italy, West Germany and Holland. Italy, too, built 245 of the final F-104S version.

Aircraft: Lockheed F-104C
Year: 1958
Type: fighter-bomber
Manufacturer: Lockheed Aircraft Co.
Engine: General Electric J79-GE-7
Power: 15,800lb (7,167kg)
Wingspan: 21ft 9in (6.62m)
Length: 54ft 8in (16.66m)
Height: 13ft 6in (4.11m)
Wing area: 196.1sq ft (18.21m²)
Max take-off weight: 27,853lb (12,634kg)
Empty weight: 12,760lb (5,788kg)
Max speed: 1,150mph at 50,000ft (1,851km/h at 15,240m)
Service ceiling: 58,000ft (17,678m)
Range: 850mi (1,368km)
Crew: 1
Load-armament: 1x20 mm cannon; 2,000lb (907kg)

Republic F-105

Designed in 1954 as an all-weather supersonic attack aircraft, capable of carrying nuclear or conventional armaments, the F-105 was the culmination of the series of warplanes built by Republic since the early days of World War II. The prototype had its first flight on October 22, 1955, and production began with 71 F-105B, which started to be delivered in May 1958. On June 9, 1959 the prototype appeared of the second, more powerful version, the F-105D, with a more powerful engine and improved electronic equipment. This went into service in 1961, with 610 machines built. The last variant was the two-seater, advanced trainer F-105F, delivered from 1963 (143 planes). During its busy operational career, 350 F-105Ds were continuously strengthened and modernized, particularly in the electronic field.

Aircraft: Republic F-105D
Year: 1959
Type: fighter-bomber
Manufacturer: Republic Aviation Corp.
Engine: Pratt & Whitney J75-P-19W
Power: 26,500lb (12,020kg)
Wingspan: 34ft 11in (10.64m)
Length: 64ft 3in (19.58m)
Height: 19ft 8in (5.99m)
Wing area: 385sq ft (35.76m²)
Max take-off weight: 52,546lb (23,835kg)
Empty weight: 27,500lb (12,474kg)
Max speed: 1,372mph at 36,000ft (2,208km/h at 10,973m)
Service ceiling: 32,100ft (9,784m)
Range: 900mi (1,448km)
Crew: 1
Load-armament: 1x20mm cannon; 14,000lb (6,350kg)

Republic F-105D Thunderchief of the 333rd TFS.

◁ Lockheed F-104C Starfighter of the 435th Tactical Fighter Squadron.

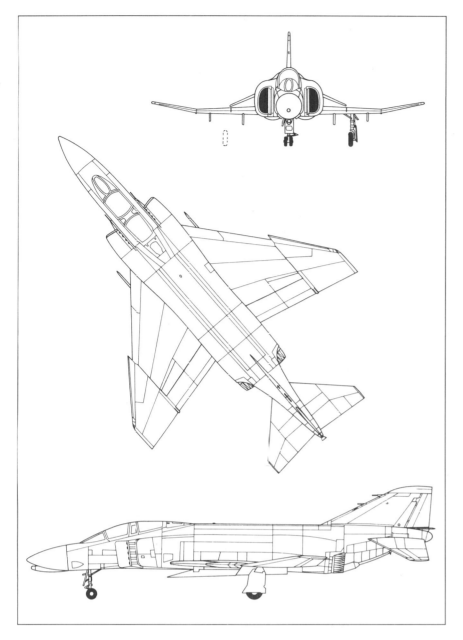

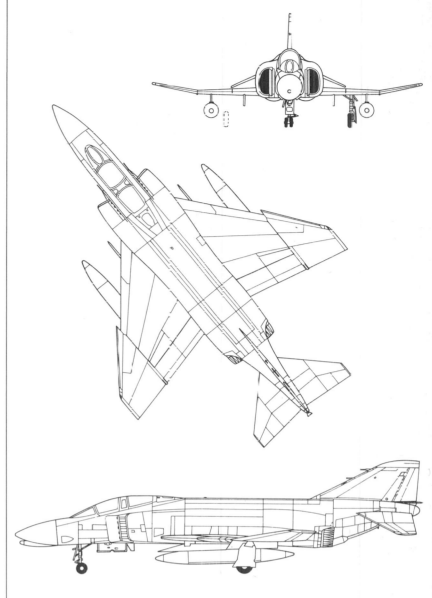

McDonnell F-4B

Unanimously recognized as the best fighter-bomber ever built, the F-4 Phantom II was designed in 1953 with a view to providing the US Navy with an all-weather supersonic twin-jet capable of combining speed, maneuverability, bombload capacity, weight and power. No easy task, but the McDonnell designers succeeded brilliantly; when on May 27, 1958, the first prototype (F4H-1) took to the air, its qualities were so obvious that the US Navy chose it in preference to its direct rival, the LTV F8U-3 Crusader III, ordering its mass-production. The first basic version, designed for shipboard use by the US Navy and the Marines, was the F-4B (first flight March 25, 1961), and 649 of these planes were delivered up to 1967. In addition to the many variants adopted by the USAF, the US Navy took 522 of a second version, the F-4J Phantom II (first flight May 1966). In Vietnam the F-4s were first sent into action from the aircraft carrier USS Constellation on August 5, 1964.

Aircraft: McDonnell F-4B
Year: 1961
Type: fighter-bomber
Manufacturer: McDonnell Aircraft Corp.
Engine: 2 x General Electric J79-GE-8A
Power: 17,000lb (7,711kg)
Wingspan: 38ft 5in (11.70m)
Length: 58ft 3 $^3/_4$in (17.78m)
Height: 16ft 3in (4.95m)
Wing area: 530sq ft (49.23m²)
Max take-off weight: 44,600lb (20,231kg)
Empty weight: 28,000lb (12,701kg)
Max speed: 1,485mph at 48,000ft (2,390km/h at 14,630m)
Service ceiling: 62,000ft (18,898m)
Range: 400mi (644km)
Crew: 2
Load-armament: 6-8 missiles; 16,000lb (7,275 kg)

McDonnell F-4C

The Phantom II, among its other achievements, was the first fighter designed specifically for shipboard use to be adopted as well by the USAF. Its successful «ground» career began on March 30, 1962 (after a single F-4B had proved itself far superior to a Convair F-106A) when the USAF placed an order for an air-superiority and tactical support version. The prototype of this, the F-4C, took to the air on May 27, 1963, and 583 were eventually built. Then followed 503 RF-4C photo-reconnaissance planes (delivery commencing June 1964), 825 F-4Ds (first flight December 7, 1965), and about 1,500 F-4Es (first flight June 30, 1967), of which almost one-third were exported. Overall production ended in October 1979, by which time over 5,100 Phantom IIs had been built in the USA and 140 under license in Japan. The F-4 flew under the USAF insignia for practically the entire period of the Vietnam War. From 1975 the Pantom IIs were gradually replaced by F-14s, and al-

Aircraft: McDonnell F-4C
Year: 1963
Type: fighter-bomber
Manufacturer: McDonnell Aircraft Corp.
Engine: 2 x General Electric J79-GE-15
Power: 17,000lb (7,711kg)
Wingspan: 38ft 5in (11.70m)
Length: 58ft 3 $^3/_4$in (17.78m)
Height: 16ft 3in (4.95m)
Wing area: 530sq ft (49.23m²)
Max take-off weight: 51,441lb (23,334kg)
Empty weight: 28,496lb (12,926kg)
Max speed: 1,433mph at 40,000ft (2,306km/h at 12,192m)
Service ceiling: 56,100ft (17,099m)
Range: 538mi (866km)
Crew: 2
Load-armament: 4 missiles; 16,000lb (7,275kg)

though mainly consigned to reserve units, they are still used for front line duty in many countries, and will probably continue in this role until the late 1990s.

McDonnell F-4B of the 111th Fighter Squadron.

McDonnell F-4C of the 557th TFS.

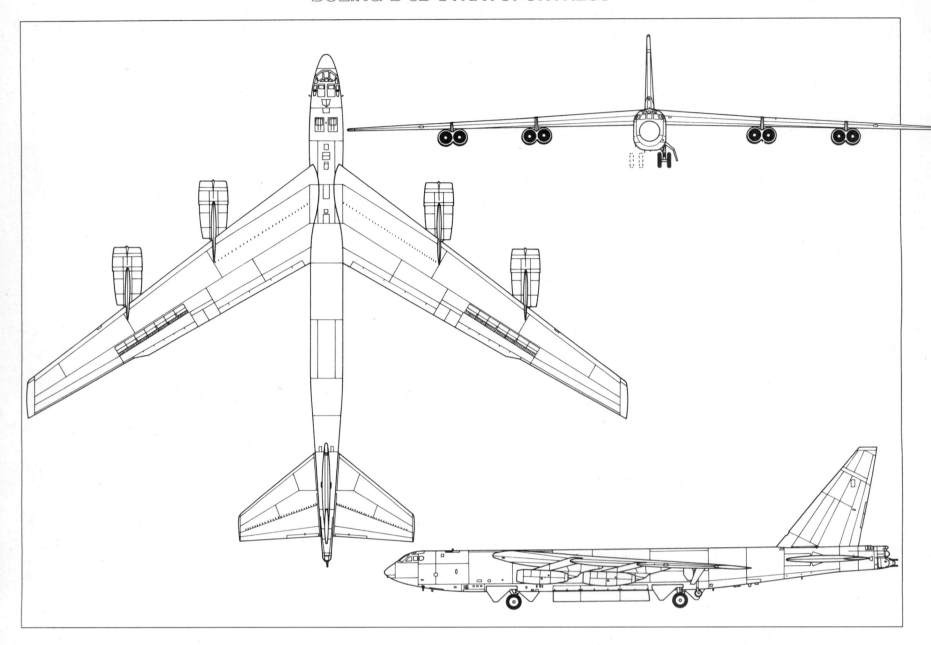

Boeing B-52

The Boeing B-52 dates back to April 1945, when consideration began to be given to developing a new strategic bomber destined to succeed the B-36, then in the advanced planning stage. The design was revised several times and it was only toward the end of 1948 that the final design was approved. The first flight of the prototype took place on April 15, 1952, and the first of the three initial B-52A series models appeared on August 5, 1954. These aircraft became operational only in November 1957, after a long series of evaluations. In the meantime sixteen examples of the photo-reconnaissance version RB-52B had already been built; in addition to these were 38 B-52Bs, fifteen of which could be converted for reconnaissance duties. The 35 B-52Cs of 1956 were heavier, with greater fuel capacity, while the D version (170 machines), which appeared on June 4, 1956, was designed exclusively as a bomber. This was followed from 1957 by 100 B-57Es, with improved electronics, and from 1958 by 88 planes in the F version, with more powerful engines. The last variants were the G and H versions (bringing total production to 744 planes), achieving the utmost in performance and fighting potential.

Boeing B-52F of the 320th Bomb Wing.

Boeing B-52D Stratofortress at Andersen.

Aircraft: Boeing B-52D
Year: 1956
Type: bomber
Manufacturer: Boeing Aircraft Co.
Engine: 8 x Pratt & Whitney J57-P-29W
Power: 12,100lb (5,490kg)
Wingspan: 185ft 0in (56.38m)
Length: 157ft 7in (48.03m)
Height: 48ft 4 $\frac{1}{2}$in (12.40m)
Wing area: 4,000sq ft (371.6m²)
Max take-off weight: 450,000lb (204,120kg)
Empty weight: 171,000lb (77,567kg)
Max speed: 630mph at 24,000ft (1,014km/h at 7,315m)
Service ceiling: 55,000ft (16,675m)
Range: 6,200mi (9,978 km)
Crew: 6
Load-armament: 4 tail guns; 70,000lb (31,752kg)

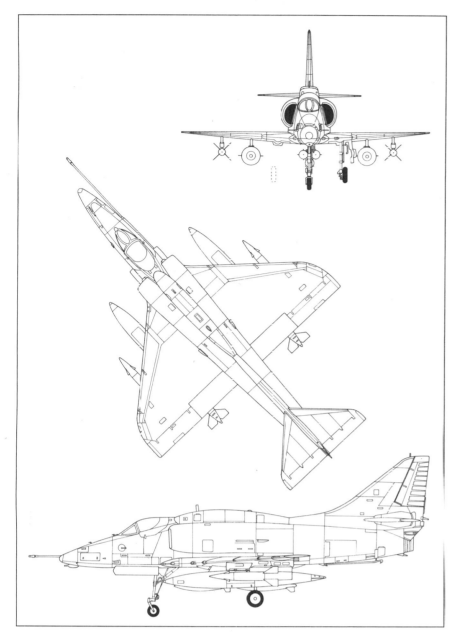

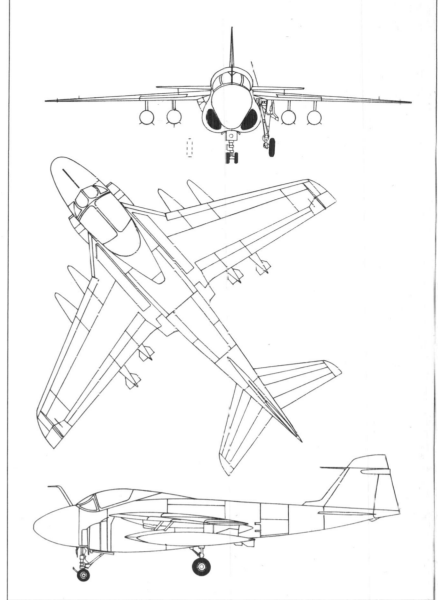

Douglas A-4

Almost 3,000 of these aircraft were built in 25 years between 1954 and 1979. This figure alone gives some idea of the enormous success (both at home and abroad) of the Douglas A-4 Skyhawk, one of the most effective carrier-based attack planes, operational from 1956 and used in all the major engagements of the 1960s and 1970s. The program was launched when the US Navy decided it wanted a modern replacement of the Skyraider. The requirements were most exacting, but Ed Heinemann, chief designer at Douglas, managed to satisfy them all, proposing a plane whose maximum weight at take-off was practically half of the Navy specification weight, giving it increased payload capacity. The first prototype flew on June 22, 1954 and the first production model of the A4D-1 (designated A-4A after 1962) on August 14. There were many subsequent variants and subseries derived from this

model. The prototype of the A-4F version appeared on August 31, 1966, and deliveries of the 146 planes ordered by the US Navy took place between June 1967 and June 1968.

Aircraft: Douglas A-4F
Year: 1966
Type: attack
Manufacturer: Douglas Aircraft Co.
Engine: Pratt & Whitney J52-P-8A
Power: 9,300lb (4,218kg)
Wingspan: 27ft 6in (8.38m)
Length: 40ft 3 $^{1}/_{4}$in (12.27m)
Height: 15ft 0in (4.57m)
Wing area: 260sq ft (24.16m²)
Max take-off weight: 24,500lb (11,113kg)
Empty weight: 10,000lb (4,535kg)
Max speed at sea level: 675mph (1,086km/h)
Service ceiling: 49,000ft (14,935m)
Range: 2,000mi (3,200km)
Crew: 1
Load-armament: 2x20mm cannon; 10,000lb (4,500kg)

Grumman A-6

One of the finest all-weather attack planes of the US Navy, the Grumman A-6 Intruder is still in front line service, in increasingly up-to-date versions, particularly as regards electronics and armament. In 1957 the Grumman emerged as the winner in competition with eleven other rival designs entered by eight companies. The first prototype flew on April 19, 1960, and the first A-6A of the series went into service on February 1, 1963; 488 of them were built by the end of 1969. The second basic version was designed exclusively for electronic warfare. Named EA-6B (and stemming from an initial subseries of 27 models), the prototype appeared on May 25, 1968; the fuselage and cabin were modified to accommodate four crew members as well as a highly sophisticated electronics system, so that it was soon recognized as the best carrier-based aircraft for this type of warfare. The fi-

Aircraft: Grumman A-6A
Year: 1960
Type: attack
Manufacturer: Grumman Aircraft Engineering Corp.
Engine: 2 x Pratt & Whitney J52-P-8A
Power: 9,300lb (4,218kg)
Wingspan: 53ft 0in (16.15m)
Length: 54ft 7in (16.64m)
Height: 15ft 7in (4.75m)
Wing area: 529sq ft (49.15m²)
Max take-off weight: 60,626lb (27,500kg)
Empty weight: 25,684lb (11,650kg)
Max speed at sea level: 685mph (1,102km/h)
Service ceiling: 41,660ft (12,700m)
Range: 1,920mi (3,090km)
Crew: 2
Load-armament: 15,000lb (6,804kg)

nal variant of the Intruder was the A-6E, dating from February 1970, even further improved and more powerful.

Douglas A-4 Skyhawk of the 93th Attack Squadron.

Grumman A-6 Intruder on the USS Constellation.

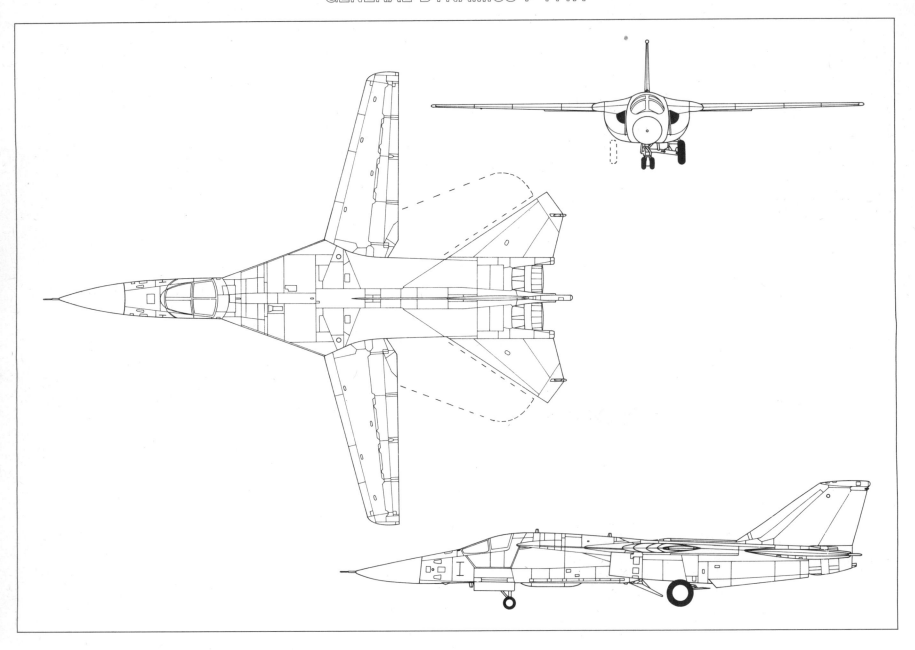

General Dynamics F-111

The first warplane with variable geometric wings to be mass produced, the F-111 was built by General Dynamics at the beginning of the 1960s against an initial contract for delivery of eighteen planes to the USAF and five to the US Navy. Although the prototype had already flown on December 21, 1964, subsequent phases of development were fraught with difficulties, leading to the abandonment of the program by the US Navy. Production, therefore, was reserved for the USAF, which took delivery of the first machines in 1968. The first variant was

the F-111A, and after seventeen pre-production models, 141 of these planes were built. Then followed 76 FB-111As, with more powerful engines, larger wings and more sophisticated electronics, which went into service in 1969. Successive variants were the E (94 planes in service from September 1970); the D for tactical support (96 machines, operational from October 1971); and the F, the final version which appeared as prototype in May 1973 and of which 106 were made, the last in November 1976. Total production of General Dynamics F-111 was 562 aircraft, including 24 F-111Cs which were exported to Australia.

General Dynamics F-111A of the 429th TFS.

Aircraft: General Dynamics F-111A
Year: 1964
Type: fighter-bomber
Manufacturer: General Dynamics
Engine: 2 x Pratt & Whitney TF30-P-1
Power: 18,000lb (8,165kg)
Wingspan: 63ft 0in (19.20m)
Length: 73ft 5 $\frac{1}{2}$in (22.40m)
Height: 17ft $\frac{1}{2}$in (5.18m)
Wing area: 525sq ft (48.77m²)

Max take-off weight: 98,850lb (44,838kg)
Empty weight: 46,172lb (20,944kg)
Max speed: 1,435mph at 53,450ft (2,338km/h at 16,292m)
Service ceiling: 56,650ft (17,267m)
Range: 1,330mi (2,140km)
Crew: 2
Load-armament: 1x20mm cannon; 30,000lb (13,608kg)

General Dynamics F-111A of the 428th Tactical Fighter Squadron.

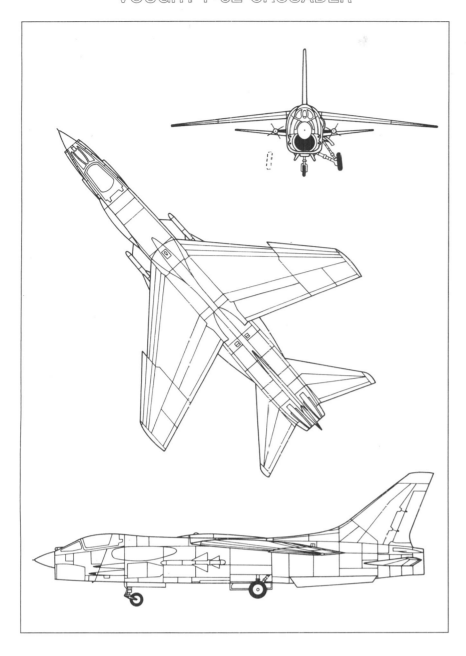

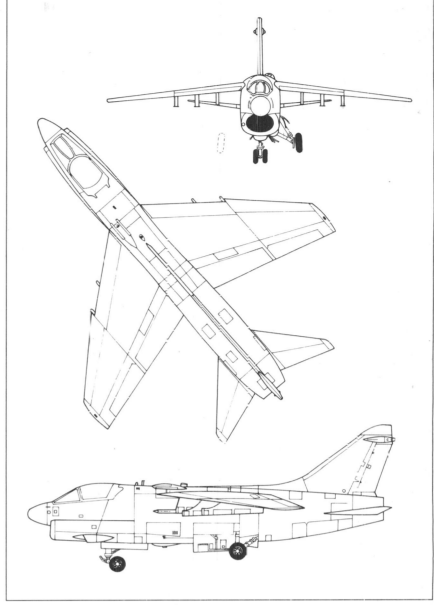

Vought F-8

First daytime supersonic carrier-based interceptor, the Crusader was conceived in 1952, built to the order of the US Navy. The prototype flew on March 25, 1955, successfully testing the original solution of a wing of variable incidence, designed to reduce the velocity and improve landing capability. Production started soon afterward and, up to 1965, 1,259 machines of various series were built, increasingly powerful and up-to-date. The principal types were the F-8A interceptor (first flight September 30, 1955, with 318 machines), which went into service in March 1957; the F-8C air-superiority (187 machines, first flight August 20, 1958); and the F-8D all-weather fighter (152 planes, first flight February 16, 1960). The final version was the F-8E (first flight June 26, 1964), also designed as an all-weather fighter, of which 286 were built. The Crusader remained in front line service until the late 1970s, thanks to a massive modernization program starting in 1966, involving 375 planes from all series: 136 F-8Es,

in particular, were transformed into F-8Js, the modifications applying mainly to certain structural components, the wings and the electronics. In addition to service in the US Navy, F-8Es were taken by the French Aéronavale.

Aircraft: Vought F-8E
Year: 1964
Type: fighter
Manufacturer: Chance Vought Inc.
Engine: Pratt & Whitney J57-P-20A
Power: 18,000lb (8,165 kg)
Wingspan: 35ft 8in (10.87m)
Length: 54ft 3in (16.53m)
Height: 15ft 9in (4.80m)
Wing area: 375sq ft (34.83m²)
Max take-off weight: 34,000lb (15,422kg)
Max speed: 1,120mph at 40,000ft (1,802km/h at 12,192m)
Service ceiling: 58,000ft (17,678m)
Crew: 1
Load-armament: 4x20mm cannon; 5,000lb (2,267kg)

Vought A-7

Planned in 1963 as the successor to the Douglas A-4 Skyhawk for the US Navy and Marines, the A-7 Corsair II proved so effective that it was also chosen by the USAF to replace the F-100s and F-105s. The prototype flew for the first time on September 27, 1965, and production commenced on 199 of the A-7A version. This was followed by the more powerful A-7B (196 machines, first flight February 6, 1968). The variant designed for the USAF was the A-7D, which took off on April 5 furnished with a different kind of engine and modified both with regard to armament and electronics. Deliveries of the 459 models of the A-7D which had been ordered took place from September 1970 to December 1976. In 1969 a new version was produced for the US Navy, the A-7E, which became the principal type built (after the first 67 machines, known as A-7C), with 529 planes up to March 1981. Among minor variants were the TA-7C and the A-7K, two-seater trainers for possible operational use by the US

Navy and the US Air National Guard; and the A-7H and A-7P for Greece and Portugal. It was much used in Vietnam; the first A-7As received their baptism of fire on December 4, 1967.

Aircraft: Vought A-7D
Year: 1968
Type: attack
Manufacturer: LTV Aerospace Corp.
Engine: Allison TF41-A-1
Power: 14,250lb (6,465kg)
Wingspan: 38ft 9in (11.80m)
Length: 46ft 1 ½in (14.06m)
Height: 16ft 0in (4.88m)
Wing area: 375sq ft (34.83m²)
Max take-off weight: 42,000lb (19,051kg)
Empty weight: 19,781lb (8,972kg)
Max speed at sea level: 698mph (1,123km/h)
Range: 951mi (1,762km)
Crew: 1
Load-armament: 1x20mm cannon; 15,000lb (6,804kg)

Vought F-8E Crusader at Da Nang, March 1967.

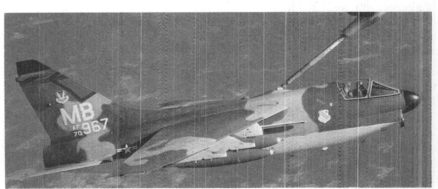

Vought A-7D of the 354th TFW.

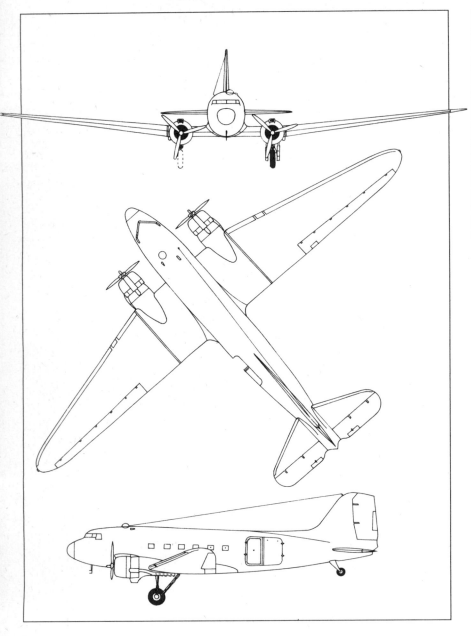

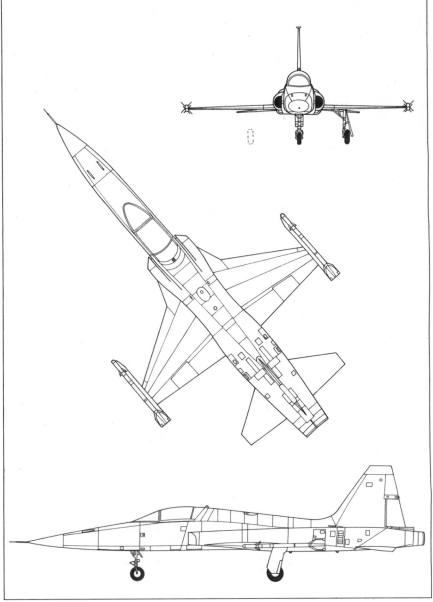

Douglas AC-47

One of the planes to see service in the Vietnam War was the immortal Douglas DC-3, in its C-47 military version. Although first made as far back as 1935 (total production of the DC-3 was over 13,000 machines, almost 11,000 of them built in the United States, and of these, 10,123 in the C-47 military version) the trusty Dakota proved its worth in the unaccustomed role of «flying gunship.»

The first prototype of the Douglas DC-3, derived directly from the DC-1 and DC-2, of which it was the final version, took to the air on December 17, 1935. The DC-3, like its predecessor, was built at the request of American Airlines, who advised Douglas to enlarge the fuselage of the DC-2 so as to accommodate fourteen berths to be used on long coast-to-coast flights by night, so replacing the service provided by the obsolete three-engined Fokker and

Aircraft: Douglas AC-47B
Year: 1941
Type: attack
Manufacturer: Douglas Aircraft Co.
Engine: 2 x Pratt & Whitney R-1830-92, rad., 9 cyl., air cooled
Power: 1200hp
Wingspan: 95ft 0in (28.96m)
Length: 64ft 5 $^1/_2$in (19.63m)
Height: 49ft 2in (5.20m)
Wing area: 987sq ft (91.70m²)
Max take-off weight: 26,025lb (11,805kg)
Empty weight: 16,986lb (7,705kg)
Max speed: 198mph at 7,545ft (368km/h at 2,300m)
Service ceiling: 23,215ft (7,075m)
Range: 1,296mi (2,400km)
Crew: 5
Load-armament: 3x7.62mm GE minigun

Curtiss Condor biplanes. In practical terms the outcome was a bigger, sturdier and more powerful version of the DC-2, originally known as the DST (Douglas Sleeper Transport). Its military career began in 1941. The DC-3 was soon being sold all over the world: licenses to build were also given to the Japaneses, who produced 450 machines in the L2D version, and to the Russians, who built more than 2,000, as the Lisunov Li. 2. During the war the DC-3 was the principal allied transport aircraft. The basic versions in service with the USAF were the transport C-47 Skytrain and the passengers C-53 Skytrooper. The US Navy version was designated the R4D, and the name Dakota was given to the British RAF DC-3.

Northrop F-5

Planned at the start of 1955, the Northrop F-5 has become one of the most popular tactical fighters of the 1980s. Designed primarily for export, the aircraft's success is mainly due to its simplicity, lightness of weight and low cost, which, united with excellent performance and good armament, make it a worthy rival of bigger, stronger and more sophisticated planes. The F-5 prototype flew on July 31, 1959, and production was soon in full swing. About 1,300 of the early A and B variants (single - and two-seaters respectively, first flights July 31, 1963 and February 24, 1964) were built and sold to some twenty countries. Toward the end of the 1960s Northrop brought out the improved and more powerful F-5E Tiger (first flight August 11, 1972), which sold as successfully as its predecessors. The F-5As went into service with the USAF in August

Aircraft: Northrop F-5A
Year: 1963
Type: fighter
Manufacturer: Northrop Corp.
Engine: 2 x General Electric J85-GE-13
Power: 4,080lb (1,851kg)
Wingspan: 25ft 10in (7.87m)
Length: 47ft 2in (14.37m)
Height: 13ft 2in (4.01m)
Wing area: 170sq ft (15.79m²)
Max take-off weight: 20,677lb (9,379kg)
Empty weight: 8,085lb (3,667kg)
Max speed: 925mph at 36,089ft (1,489km/h at 11,000m)
Service ceiling: 50,500ft (15,392m)
Range: 558mi (898km)
Crew: 1
Load-armament: 2x20mm cannon; 6,200lb (2,812kg)

1964, and in October 1965 a few machines were sent experimentally to Vietnam.

Douglas AC-47 with three rotary 7.62mm miniguns.

Northrop F-5C and F-5A.

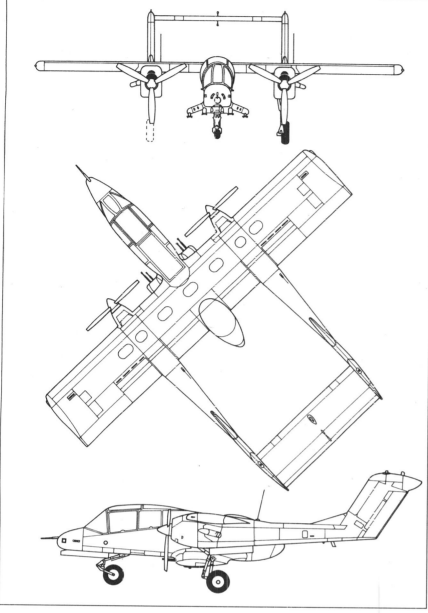

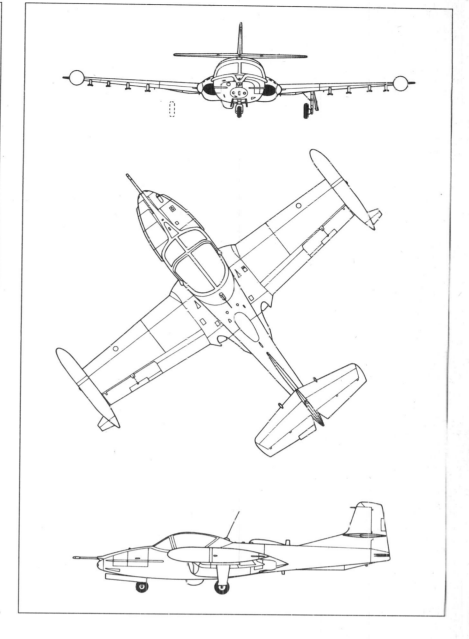

Rockwell OV-10

More like a helicopter than an airplane, the Rockwell Bronco was the ideal example of a tactical reconnaissance plane designed to work in close cooperation with ground forces. Planning of the OV-10 began in 1962 with orders from the USAF, the US Navy and the US Marine Corps, the objective being to produce an armed reconnaissance plane specialized in anti-guerilla operations. The definitive prototype flew on August 15, 1966, and the first OV-10A production model appeared on August 6, 1967. Up to April 1969, 271 of these planes were built, 114 for the Marines and 157 for the USAF. Production was increased to satisfy export orders to West Germany, Venezuela and Indonesia. In Vietnam the Bronco was operational from 1967.

Aircraft: Rockwell OV-10A
Year: 1967
Type: observation
Manufacturer: Rockwell International
Engine: 2 x Garrett AiResearch T76-G-10/12
Power: 715shp
Wingspan: 40ft 0in (12.19m)
Length: 39ft 9in (12.12m)
Height: 15ft 1in (4.62m)
Wing area: 291sq ft (27.03m²)
Max take-off weight: 14,444lb (6,550kg)
Empty weight: 7,190lb (3,260kg)
Max speed: 244mph at 10,000ft (452km/h at 3,048m)
Service ceiling: 18,000ft (5,486m)
Range: 165mi (306km)
Crew: 2
Load-armament: 4x7.62mm machine guns; 4,600lb (2,086kg)

Cessna A-37

In the first part of the 1950s Cessna, too, went in for military production, designing the first jet trainer for the USAF, the T-37. From this machine (of which 1,268 models of three basic versions were built between 1955 and 1977), an efficient attack plane was derived in 1963, the A-37, which was also successfully exported. The prototype flew on October 22 and the first 39 A-37As were produced by direct conversion of other T-37Bs. The definitive version was the A-37B, which first appeared in September 1967 and of which 577 were built. Most of these went to the USAF, but the aircraft was also sent to South Vietnam, Chile, Peru and Guatemala.

Aircraft: Cessna A-37B
Year: 1967
Type: attack
Manufacturer: Cessna Aircraft Co.
Engine: 2 x General Electric J85-GE-17A
Power: 2,850lb (1,293kg)
Wingspan: 35ft 10 ¹/₂in (10.93m)
Length: 29ft 3in (8.92m)
Height: 8ft 10 ¹/₂in (2.70m)
Wing area: 183.9sq ft (17.09m²)
Max take-off weight: 15,000lb (6,804kg)
Empty weight: 5,873lb (2,670kg)
Max speed: 478mph at 15,000ft (769km/h at 4,572m)
Service ceiling: 32,100ft (9,785m)
Range: 450mi (724km)
Crew: 2
Load-armament: 1x7.62mm minigun; 5,400lb (2,450kg)

Rockwell OV-10A Bronco.

Cessna A-37 Dragonfly.

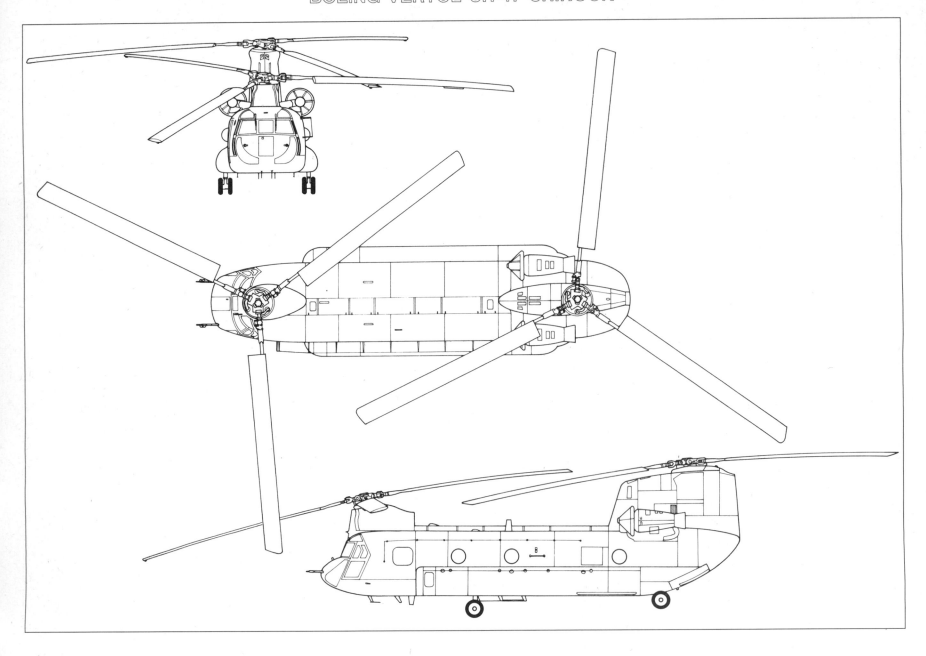

Boeing CH-47

Unreplaceable in Vietnam by reason of its exceptional ability of operating brilliantly in all conditions, the CH-47 was developed in 1965 on a specification from the US Army for a medium-size, all-weather transport helicopter with turbine engines. The first prototype took off on September 21, 1961, and production commenced very soon with the CH-47A version known as the Chinook. In October 1966 the first prototype of a second variant appeared, with more powerful engines and rotors, and deliveries began in May 1967. Then followed (with the first flight on October 14, 1967) the C version, with further improvements to engines, transmission and fuel capacity. Deliveries started in the spring of 1968 and in September these Chinooks were sent to Vietnam. At the beginning of 1969 no less than 270 CH-47s were operating at the front. These helicopters proved themselves especially valuable in rescue work and in recovering damaged aircraft.

Boeing Vertol CH-47A in 1967.

Boeing Vertol CH-47A Chinook.

Helicopter: Boeing ACH-47A	**Height:** 18ft 7in (5.67m)
Year: 1966	**Max take-off weight:** 40,000lb
Type: transport	(18,144kg)
Manufacturer: Vertol Division of	**Empty weight:** 19,555lb (8,870kg)
Boeing Co.	**Max speed:** 144mph (232km/h)
Engine: 2 x Lycoming T55-L-7C	**Service ceiling:** 9,000ft (2,745m)
Power: 2850shp	**Range:** 1,250mi (2,021km)
Rotor diameter: 60ft 0in (18.29m)	**Crew:** 2
Fuselage length: 51ft 0in (15.54m)	**Load-armament:** 33-44 troops or
Overall length: 99ft 0in (30.18m)	19,555lb max (8,870kg)

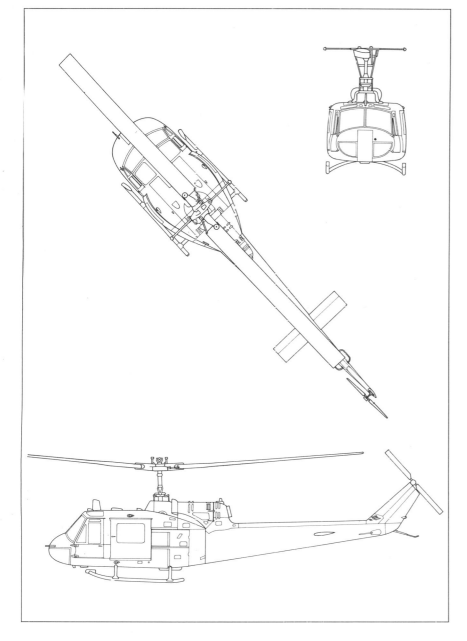

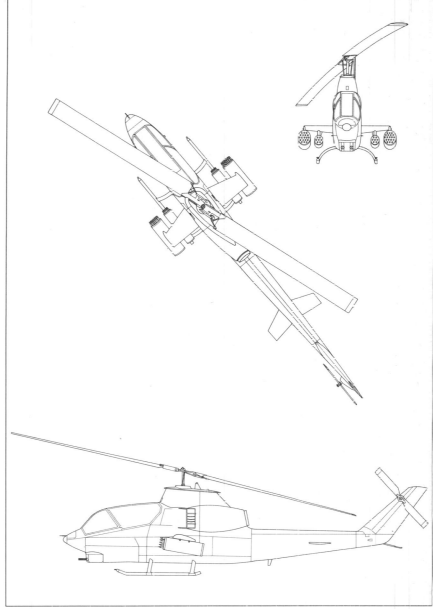

Bell UH-1

One of the all-purpose helicopters of the Vietnam War, the Bell Model 204 (better known as the Huey, from the initial letters of the first HU-1 production version) was planned in 1954 at the request of the US Army. They needed a multipurpose helicopter which could be used particularly for battlefield evacuation and for navigational training. The first of three prototypes took off on October 20, 1956, and was followed by nine pre-series models. The first version was the UH-1A (known as the Iroquois), and deliveries of this began on June 30, 1959, being completed in March 1961. The second was the UH-1B, more powerful and better adapted to carry a wide range of weaponry. Planning of this variant started in June 1959, and the first prototype flew in 1960, with deliveries to units commencing in March 1961. The UH-1Bs became operational in Vietnam in the fall of 1962.

Helicopter: Bell UH-1B
Year: 1960
Type: utility/transport
Manufacturer: Bell Helicopter Co.
Engine: Lycoming T53-L-11
Power: 1100shp
Rotor diameter: 44ft 0in (13.41m)
Fuselage length: 38ft 5in (11.70m)
Overall length: 53ft 0in (16.15m)
Height: 14ft 0in (4.42m)
Max take-off weight: 9,500lb (4,310kg)
Empty weight: 5,055lb (2,293kg)
Max speed: 138mph (222km/h)
Service ceiling: 21,000ft (6,400m)
Range: 286mi (460km)
Crew: 2
Load-armament: 7 troops or 3,000lb (1,360kg)

Bell AH-1

Developed in the first half of 1965, the Bell HueyCobra was the logical armed successor to the UH-1B and more powerful UH-1C. The new machine retained virtually the same type of engine and transmission, but the fuselage was more streamlined, designed to achieve maximum speed, maximum load-armament and better protection for the crew. The first AH-1G (Model 209) took off on September 7, 1965, and a contract for 110 initial production machines was signed on April 13, 1966. By October 1968 there had been orders for 838 machines, with another 170 ordered in January 1970.

Helicopter: Bell AH-1G
Year: 1965
Type: combat
Manufacturer: Bell Helicopter Co.
Engine: Lycoming T53-L-13
Power: 1100shp
Rotor diameter: 44ft 0in (13.41m)
Fuselage length: 44ft 5in (13.54m)
Overall length: 52ft 11 $^1/_2$in (16.14m)
Height: 13ft 5 $^1/_2$in (4.10m)
Max take-off weight: 9,500lb (4,309kg)
Empty weight: 6,096lb (2,765kg)
Max speed: 219mph (352km/h)
Service ceiling: 12,700ft (3,870m)
Range: 336mi (622km)
Crew: 2
Load-armament: 1x20mm minigun; rockets; missiles

Bell UH-1D of the VNAF.

Bell AH-1G HueyCobra of the 1st Air Cavalry Division.

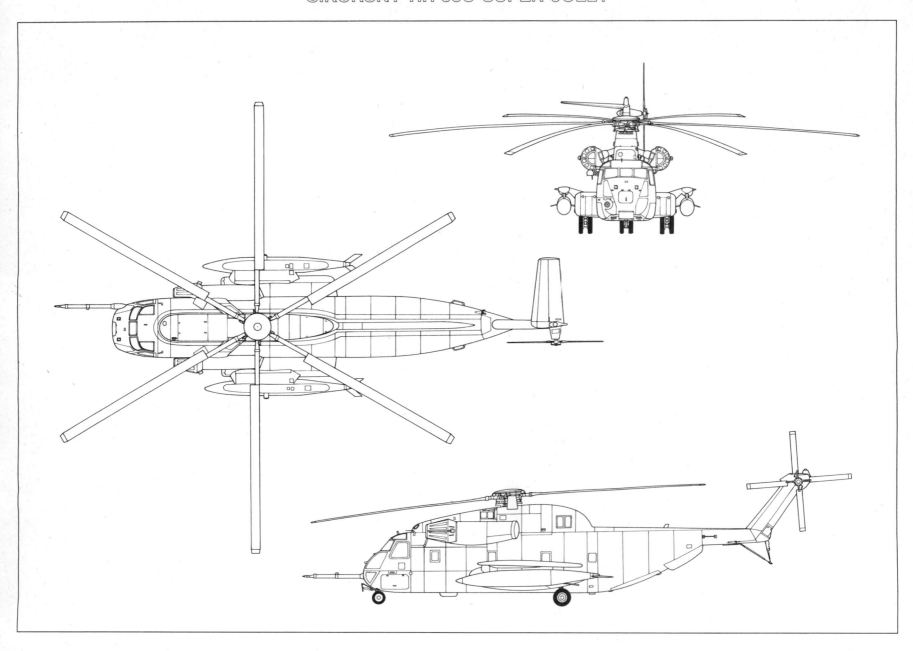

Sikorsky HH-53C

As part of the remarkable family of heavy S-61 and S-64 helicopters, Sikorsky planned a new and more powerful series of machines in the 1960s, the S-65. The first model was the CH-53A, designed in 1962 for the Marine Corps, and the second was the HH-53B, similar but built for the USAF, eight of them being ordered for the Aerospace Rescue and Recovery Service. On August 30, 1968 Sikorsky delivered the first model of a more powerful version, the HH-53C, with improvements to the engines, equipment and payload capacity. Although originally intended for the recovery of spacecraft, the HH-53C was chiefly used in Vietnam, where it was nicknamed the Super Jolly and performed excellently. Together with the CH-53A of the Marines, it was mainly utilized for rescue and recovery missions. In all, 44 machines were built before production was switched to other more powerful and improved versions.

Helicopter: Sikorsky HH-53C
Year: 1968
Type: transport
Manufacturer: Sikorsky Aircraft (Division of United Aircraft Co.)
Engine: 2 x General Electric T64-GE-7
Power: 3435shp
Rotor diameter: 72ft 3in (22.02m)
Fuselage length: 67ft 2in (20.47m)

Overall length: 88ft 3in (26.90m)
Height: 24ft 11in (7.60m)
Max take-off weight: 37,466lb (16,994kg)
Empty weight: 23,257lb (10,549kg)
Max speed: 196mph (315km/h)
Service ceiling: 20,400ft (6,220m)
Range: 540mi (869km)
Crew: 3
Load-armament: 38 troops

Sikorksy HH-3E of the 37th Air Rescue Squadron.